A NATURALIST'S GUIDE TO THE

FROGS
OF
AUSTRALIA

Scott Eipper & Peter Rowland

T0003393

JOHN BEAUFOY PUBLISHING

DEDICATION – *To the memory of Aaron Payne. Thanks for the assistance in the first edition and helping make the publication what it is today.*

This edition first published in the United Kingdom and Australia in 2023 by John Beaufoy Publishing Ltd
11 Blenheim Court, 316 Woodstock Road, Oxford OX2 7NS, England
www.johnbeaufoy.com

10 9 8 7 6 5 4 3 2 1

Photo captions and credits
Front cover: *main:* Jungguy Tree Frog © Scott Eipper; bottom row, *left to right:* Northern Ornate Nursery Frog, Giant Burrowing Frog, Stuttering Frog, all © Scott Eipper.
Back cover: Water Frog © Scott Eipper
Title page: Green Tree Frog © Tyese Eipper
Contents page: Orange-thighed Frog © Scott Eipper

ISBN 978-1-913679-35-4

Edited by Krystyna Mayer
Designed by Gulmohur Press, New Delhi

Printed and bound in Malaysia by Times Offset (M) Sdn. Bhd.

·CONTENTS·

INTRODUCTION

Frogs (and toads), of the family Anura, belong to the large class of backboned animals known as amphibians (Amphibia), which also includes the salamanders and newts (family Caudata) and worm-like caecilians (family Gymnophiona), although frogs are by far the largest family, making up almost 90 per cent of the group. Amphibians were the first vertebrates to colonize the land, towards the end of the Devonian period about 370 million years ago (mya), and most still require water for the larval stage of their life cycle. Their ancestors most resembled modern-day coelacanths, but the modern-day amphibians are vastly different from them. Amphibians are found on every continent of the world with the exception of Antarctica. They are all ectothermic ('cold-blooded'), regulating their body temperature using outside sources, all have a three-chambered heart, and many absorb oxygen through the skin, which is kept moist with mucous secretions.

Australia does not have any naturally occuring salamanders, newts or caecilians, although the Axolotl, a species of salamander, is commonly kept by enthusiasts, and escaped individuals occur from time to time. The Smooth Newt *Lissotriton vulgaris* has become established in Australia and is now considered part of the Australian herpetofauna.

Australia is home to 247 species and ten subspecies of frog, about 95 per cent of which are endemic. Australia is also home to an established introduced toad, the Cane Toad (native to South and Central America), which was introduced in 1935 from Hawaii as a biological control for French's and Greyback Cane Beetles, which were causing significant economic damage to sugar-cane farms. Sadly, the Cane Toad not only did not control the

Smooth Newt

cane beetles, but it quickly adapted to the environment in Australia's north. Breeding rapidly (and laying up to 35,000 eggs at a time), with a voracious appetite and no known predators in its new home, it has spread to almost every state and territory, although it is absent from ACT, Vic and Tas, and considered as a vagrant to SA. It is also equipped with a highly toxic combination of poisons in the large parotoid glands on its neck, making it lethal to most animals that ingest it as a potential food item. Another introduced toad, the South-east Asian Toad, has also been recorded in multiple states, but whether it has established viable populations in any areas is yet to be determined.

The number of identified frog species in Australia is constantly varying, subject to changes in current classification, the authority followed, new species discoveries and, sadly, extinction of known species. About 15 per cent of Australia's modern frog species are either extinct or at risk of extinction. Four of modern Australian frog species have been classed as extinct under the Environment Protection and Biodiversity Conservation Act. These are the Southern Gastric-brooding Frog *Rheobatrachus silus*, Northern Gastric-brooding Frog, *R. vitellinus* Sharp-snouted Frog *Taudactylus acutirostris* and Southern Day Frog *T. diurnus*. A further five species are classified as Critically Endangered, 14 as Endangered and 10 as Vulnerable. Unfortunately, the numbers of these and many other species are in decline, but perhaps most alarming is their decline in environments that are considered to be 'pristine'.

FROGS OR TOADS?

Both frogs and toads are grouped together in the order Anura. The 'true' toads of the family Bufonidae, of which there are about 600 species worldwide, are not native to Australia. Many species of true toad have the typical dry, 'warty' skin, well-developed parotoid poison glands and stocky forelimbs. They also typically lay their eggs in long strands, as opposed to the clumps of eggs laid by many frogs. There are always exceptions to the rules, however, and some species of toad have smooth, colourful, 'frog-like' skin (for example members of the South American genus *Atelopus*), while some frogs have dry, warty, 'toad-like' skin (such as members of the Australian genus *Uperoleia*). Members of the family Bufonidae have unnotched, moderately straight tips to their fingers and toes, which lack any distinct pads, fully webbed toes, and do not possess any maxillary teeth, although these characteristics can also be found to varying degrees in several frog families. The combination of these characteristics and breeding biology can, however, assist greatly in identifying the 'true' toads, especially in Australia. The well-established Cane Toad is widespread throughout tropical and subtropical Australia, and the South-east Asian Toad has been recorded as a vagrant in different parts of Australia, including NT, Qld, WA and Vic. In the laboratory, a true toad's identity can further be confirmed by its overlapping epicoracoid cartilages (part of the pectoral girdle), although by this stage the subject is no longer alive.

Cane Toad

Eungella Day Frog

Frogs as Indicators of Environmental Health

Frogs (and toads) are sensitive to changes in their environments and, as they are normally found in good numbers around freshwater ecosystems, they are regarded as good bioindicator organisms. As is the case with all fauna groups, there are a number of factors that negatively impact individual frogs and frog populations in general. They include loss of habitat from land clearing and changes to the natural flow of watercourses, introduced predators, competition from other animals (both native and non-native), environmental factors (such as climate change, increased rainfall acidity, saltwater intrusion and increased UV radiation), parasites, and degradation of the quality of the natural environment from pollutants and other chemicals. Many frog populations are confined to localized areas, in small pools that are isolated from other potential breeding areas by expanses of unsuitable habitat that cannot be naturally traversed by them. Other populations inhabit riverine ecosystems and are thus able to move more easily along these when the rivers are flowing.

Frogs absorb water, as do the tadpoles of species that produce aquatic larvae, and varying amounts of air through their skin. In doing so, however, they also absorb pollutants and other toxins (ecotoxicology), as well as harmful organisms, from their surrounding environment. Environments that support good numbers of frogs (both species and individuals) are generally considered 'healthy'. They are usually largely free of the above harmful factors, although there is always a degree of genetic abnormalities evident, even in 'healthy' environments. Environments that are polluted with significant levels of chemicals, including insecticides, herbicides, fungicides and heavy metals, are responsible for significantly increased incidences of abnormalities in frogs and tadpoles, such as

Adam Elliott

Abnormal metamorph frog from a polluted pond

hyperactivity, jaw deformities, brain damage, and changes to bone densities and the structure of the tail, limbs, toes and fingers.

One of the main modern-day pressures that many animal and plant species are under threat from is the effects of climate change – a long-term action that is influencing temperature, rainfall and wind patterns around the globe. As frogs are ectothermic, increases in the ambient temperature of their surroundings can have a dramatic affect on their mechanical, physical and biochemical functions. Climate change is manifested most visibly as altered temperature and rainfall regimes, both of which have a significant impact on the balance of an environment, especially delicate environments such as Australia's alpine areas and the high-altitude tropical rainforests. These changes have a direct impact on the resident and transient fauna and flora. Many frog species rely on environmental cues, including temperature and rainfall, from their environment for breeding, and altering these can drastically impact their ability to breed, as well as affecting the viability of eggs and the conditions required by the offspring once hatched. In some areas this may result in resident species breeding earlier, while in others it may limit the availability of suitable breeding ponds, thus increasing competition for food and other nutrients, or reduce the period of time that the ponds retain water, leaving insufficient time for some offspring to attain maturity. Certain desert species lay their eggs in wet mud during periods of sufficient rain, but the eggs will only hatch if the rain returns within an acceptable period of time and in sufficient quantity. In addition to causing breeding problems, a drier climate can affect food abundance.

Insects and other invertebrates that live and breed in moist ground litter can also decrease in abundance due to drier conditions. As a result, frogs and other fauna will have less food availabile to support sufficient numbers of individuals to allow for normal population fluctuations, and there will be increased competition with other fauna groups for whatever food there is. Drier leaf litter will also make these areas unsuitable for frog species that rely on a high level of ground moisture, and the associated cooler temperature, for shelter and protection from predators.

Increased temperatures do not only affect the amount and evaporation rates of surface water and ground moisture, climate change is also altering the chemical balance of available groundwater, including changing the pH and levels of pollutants, and increasing the presence of disease and disease-spreading organisms such as chytrid fungus, which are becoming more widespread. Chytridiomycete fungi are a diverse group, occurring within the soil and water in most environments. They are important for the healthy functioning of

African Clawed Frog

the environment and are responsible for the biodegradation of materials such as cellulose, chitin and pollens. Some, such as the chytrid fungus of the genus *Batrachochytrium*, cause significant disease. This fungus causes mass mortalities in frog populations (chytridiomycosis). The infection is spread through waterborne zoospores that infect the skin of the host, causing erosion, ulceration and sloughing of the skin – the host can become terminally ill after 10 days following exposure. Chytridiomycosis has devastated the world's frog populations, especially when populations are already under threat from other environmental and anthropogenic pressures. The arrival of the chytrid fungus in Australia is something of a mystery, but it is thought that it may have arrived in a shipment of African Clawed Frogs *Xenopus laevis* that were being used for research. From here the fungus was inadvertently transferred from researchers to wild frogs.

Seasonal flooding of the floodplains associated with major riverine systems is essential for frog reproduction and health. River systems that have been dammed or are subject to regulated water flows can have detrimental effects on the resident frog species if the water flow is too heavily restricted or absent during a species' normal breeding season. Water pools, drainage ditches and irrigation channels may also appear in areas where they were absent before, providing alternative options for species that can adapt to them. These species are likely to increase in numbers, while those less able to adapt will decline. The increase in competition resulting from this drives the numbers of the decreasing populations down at a more rapid rate, leading to insufficient numbers of breeding adults, and localized, regional or global loss of the species or subspecies.

As in all animal populations, an outbreak in numbers or strains of parasites can have a devastating impact on localized populations. The evidence of the effects can be seen where species range near the limit of the physiological range of the parasite. For example, chytrid fungus ceases reproduction above 29° C, so frogs in lowland areas in northern Qld have survived infection, while further upstream, in cooler upland locations, there has been localized population disappearance. Furthermore, there is anecdotal evidence that some populations are recovering, with chytrid-resistant individuals recovering seemingly 'lost' populations.

Pest animal species cause huge harm to frog populations. In the high country, horses, deer, pigs and cattle are largely responsible for the destruction of alpine bogs due to their hooved feet. The introduced Cane Toad is responsible for increased competition for resources. Direct predation of individuals by animals such as trout, mosquitofish and Cane Toads is also a factor. In many cases it is a cumulative total of pressure upon a species that leads to localized extinction events and in turn to species extinction.

Some species are particularly at risk of extinction due to one or more of the factors listed above, especially the complex issues arising from climate change (mainly the increased prevalence of the devastating chytrid fungus, prolonged droughts and altered breeding regimes). In particular, the Elegant Nursery Frog of north-east Qld is found in such a small and isolated area at the summit of Thornton Peak that it would be unable to move to a new area of suitable habitat if its current home were lost. The Southern Corroboree Frog of the Australian Alps has a similarly restricted range and habitat preference, and is already under threat from the impacts of increased temperatures, which have affected breeding and egg success, but its habitat is also being degraded by pollution to its waterways, infrastructure development related to the local skiing industry and hydroelectric scheme, erosion caused by feral animals (mainly horses, pigs and deer), and the spread of invasive weeds and chytrid fungus.

Pristine bog

Bog destroyed by feral animals

ADAPTATIONS TO AUSTRALIA'S DRY CLIMATE

While the majority of the world's frog species occur in moist or wet environments, with relatively high humidity, regular rainfall or access to permanent waterways, many species make arid areas, including deserts, their home. This is particularly true in Australia, where large parts of the country can go without rain for several years at a time. In order to survive in arid areas, the frogs that are found there have several behavioural and morphological adaptations. One of the key adaptations for desert-dwelling species is their ability to burrow, so that they can spend the drier months of the year underground. About a third of Australia's frog species burrow to some degree – some dig down rearwards, some forwards and some use a circular motion to move down through the soil. Burrowing frogs have short limbs, modified feet, typically with a sharpened edge to the bones of the foot (metatarsal tubercles), which assist with cutting through the soil, and a rounded body. Some species burrow deep down under the ground surface, while others dig down in the sand, soil or dense leaf litter, where moisture levels are higher and the temperature is more suitable (cooler or warmer) than that at ground level. Some burrowing species, particularly those of the deserts, stay underground for many months – sometimes over a year – at a time, waiting for the next rains to provide sufficient water for them to breed. Other species shelter in shallow burrows, under rocks or fallen timber, in deep crevices or within the dense foliage of different plants to avoid dehydration during the heat of the day, emerging during the comparatively cooler hours of the night.

The skin is responsible for almost all of a frog's water loss, so preventing water loss from happening is paramount to a frog's survival. This is crucial for species that live in arid areas, and even more so for those that stay underground for a lengthy period of time. Some burrowing species create a cocoon of several layers of loose dead skin around their body that traps the water beneath it. This is a modification of a natural process. The outer layer of skin on frogs regularly dies, a process known as cytomorphosis, and this layer of dead skin is shed (sloughed) to make way for a new layer to form underneath. Once sloughed, the old skin is eaten.

Beneath the skin, frogs have a series of loose sacs (lymph sacs), which are formed by thin connective tissue that attaches the skin to the underlying muscles. The sacs vary in size between species living in different environments. Those that live in wetter environments have larger sacs, while those from drier regions have smaller ones. The sacs are a key component of water regulation. Larger sacs expel water via the lymphatic system at a faster rate than smaller ones, thus ensuring that frogs in aquatic environments do not become too waterlogged and, conversely, the species in arid areas retain as much of the valuable moisture that they have absorbed as possible. While frogs from arid areas have smaller lymph sacs, their overall body size tends to be larger. Smaller frogs have a higher surface-area ratio and are thus more prone to dehydration than larger ones.

Desert frogs breed rapidly in response to sufficient rainfall during the breeding season (generally the warmer months), and lay their eggs in the temporary pools that form or in wet mud. The eggs that are laid in wet mud eventually hatch when the rains return. Other frogs, such as the sandhill frogs and Turtle Frog, do not require permanent water for reproduction – the young frogs hatch directly from eggs and there is no tadpole stage – although the substrate that the eggs are contained in needs to be moist.

BREEDING & LIFE CYCLES

Mating between male and female frogs can appear to be a very physical and haphazard affair, and well it may be in some populations where males far outweigh females, but females of some species appear to be quite selective in the choice of a male mate. It is most likely that the male is selected by a combination of his appearance and his call, but other factors also play a role, although the full extent of these is not fully understood. Once the female has made her selection, the male grasps her from above, wrapping his arms and legs around her body, and aligning his pelvic area with hers. The males of some species grab the female above the armpits, while others wrap their limbs around her waist. This embrace is termed 'in amplexus', and mature males develop nuptial pads on their thumbs during the breeding season as an aid to grip the female's slippery skin. Competition for available females is fierce between rival males, and several males may grapple over a single female. The breeding instinct is so strong in males that individuals have been recorded in amplexus with other males, females of other species, other objects and even dead females.

While in amplexus the female frog lays her jelly-covered eggs. Some species lay them in long strands (typical of the true toads), others in small or large clumps, while some lay individual, unattached eggs. The eggs laid by the female, and the covering she places over them to protect them, are known collectively as spawn. As soon as the eggs have been laid, the 'piggy-backing' male fertilizes them with his sperm. Interestingly, the jelly coating around the eggs not only serves to protect the eggs, but also aids fertilization. Spawn can float on the surface of water within a foam nest, float for a short period and then sink, or be submerged in water, attached to the surfaces of plants or rocks, laid in a burrow, or laid directly into or on soil or ground litter. The eggs are usually black, but this pigment is missing in species that lay their eggs in places that are not directly exposed to the sun, such as underground burrows.

Once the eggs are laid, adult frogs care for them and the hatchlings to varying degrees.

Foam nest

Some do no more than ensure that the site in which their eggs are laid is suitable for larval development, and that the environment that the young hatch into will aid their chances of survival, while others provide ongoing care to the eggs and/or the developing young. The eggs of most species are typically laid in a damp to aquatic environment, and the young emerge as free-swimming larvae (tadpoles) that spend the first stage of their lives fully dependent on their aquatic home, breathing with the aid of gills, as fish do. Tadpoles start life without any limbs, moving around their aquatic home with the aid of a tail. As they grow and develop, they undergo several stages of metamorphosis, typically first growing hindlimbs (see opposite), then forelimbs, after which the body shape and head structure change to resemble the adult form. They then lose their gills and are able to leave the water. This metamorphosis from larva to adult is complete when the tail is lost (it is reabsorbed into the body).

Kate Rowland

Tadpoles of different species exhibit only minor differences, although when species are closely related and not easily told apart by appearance, the tadpoles are almost impossible to distinguish, if at all. Obvious differences are, however, apparent in some species, especially between genera and families, although often these are limited to the mouth parts, the length of the tail in proportion to the length of the body, and the width of the fins along the tail. Perhaps the most obvious difference between tadpoles relates to whether they occupy lotic (flowing) or lentic (still) water. In species that occupy lotic water, the body is more dorsally compressed and the mouth is sucker-like, making the animal more streamlined and giving it an increased ability to anchor itself to rocks. Tadpoles are omnivorous, typically feeding on small particles of plant or animal matter, but are also capable of eating larger items, with food items either actively foraged for, or filtered from the water that passes through the gills.

Not all frog species lay eggs that produce tadpoles. About 15 per cent of all frogs lay eggs within which the larvae develop, then hatch as fully formed frogs. These eggs are larger, with more yolk inside to feed the developing young, and there are fewer eggs in each clutch, compared to those of tadpole-bearing species. The outer covering of the egg is also usually more durable.

At all stages of a young frog's life (including the egg and tadpole stages), it is vulnerable to predation from fish, birds, reptiles, mammals and other frogs, as well as other tadpoles. Most tadpole deaths, however, are due to their environment drying up before they reach the frog stage of development. This is of particular concern to species that breed in the summer months and must rely on temporary pools for breeding. Young frogs are also particularly vulnerable when they leave the water for the first time after metamorphosing into the adult form. They are not only at risk from predators, but must also quickly find moist shelter to prevent dehydration in their new environment.

Scott Eipper

Red-eyed Tree Frog tadpole

Australia was home to two frog species that were unique in their breeding behaviour. Northern and Southern Gastric-brooding Frogs are unique in the way they care for their developing young. The females of both species swallow the fertilized eggs (up to 40, but more commonly 20–30), shortly after the eggs are laid, then brood them within the stomach. During this time the female stops producing digestive juices, which is triggered by a chemical substance (prostaglandin E2) that coats the eggs and is also secreted by the tadpoles. When hatched, the tadpoles develop within the female for more than six weeks, at which point the fully formed frogs are gradually regurgitated. Interestingly, not all the eggs that the female lays make it to the frog stage of development, so perhaps some are accidentally digested, or are not swallowed in the first place. Sadly, both these species disappeared shortly after being discovered and are now classified as extinct.

Froglet absorbing tail

Froglet leaving the water for the first time

Another frog species that exhibits an unusual level of care for its young is the Hip Pocket Frog *Assa darlingtoni* of the cooler rainforests along the border of Qld and NSW. The female lays her eggs on land. They are quickly fertilized by the accompanying male, in much the same way as in many other frog species. After they are laid, however, the male sits and waits for them to hatch (about 11 days), then encourages the tadpoles to wriggle into special pouches on his flanks, in front of his rear legs. By covering himself in the egg jelly, he provides the tadpoles with the medium within which they can swim towards his hip pockets, although the pockets generally cannot hold all of the tadpoles that hatch (often only about half). The ones that do make it inside the pockets are then carried and protected by the male while they metamorphose, then emerge as frogs about 8–9 weeks after hatching.

HABITATS

Australia's landscape has evolved over the past 3,000 million years. It has been shaped by major geological changes, resulting from continental drift, changing sea levels, long-term wind and water erosion, and the more recent volcanic activity of only a few thousand years ago. Australia's climate has also become drier over the past 25 million years, leading to increasing soil aridity and altering the distribution and composition of the native vegetation, and it will continue to undergo change as the continent moves further north.

At localized levels Australia is home to many thousands of vegetation types. The following broad groups are the ones most widely referred to throughout this guide and broader published material.

RAINFORESTS Including tropical, subtropical and cool temperate rainforests. Found in the wetter climatic zones of Australia and typically characterized by dense foliage and a high diversity of plant species.

TALL OPEN FORESTS Tall trees more than 30m in height found in wetter regions of the country. Australia's tallest species and the tallest hardwood tree in the world is the Swamp Gum, or Mountain Ash, *Eucalyptus regnans*, with individuals growing to more than 100m in height.

OPEN FORESTS Trees typically 10–30m in height and widespread in eastern (including Tas), northern and southwestern Australia. They have a shrubby or grassy understorey.

LOW OPEN FORESTS Trees with an average height of 5–10m, generally found in areas that are cooler, drier, lower in nutrients, have rocky slopes or are subject to regular flooding.

WOODLAND/OPEN WOODLAND Some woodland can contain a diverse assemblage of mixed tree species, many restricted in total range. Other woodland is dominated by a single genus of plant. Open woodland has a wider spacing between trees, allowing neighbouring grassland and shrubland to invade, and forming a valuable mosaic of these different vegetation communities. Dominant woodland types include eucalypt, acacia, callitris, casuarina and melaleuca.

Rainforest

SHRUBLAND Typified by multi-stemmed shrubs, either monotypic or a broad range of shrub species. The dominant shrubland type in Australia is acacia, including mulga and gidgee, with other shrubland having a mix of grevilleas, samphires, saltbush, chenopods, banksias and emu bushes.

HEATHS Typically a mixture of species, many with a mature height of 1m or less, with dense canopies. Associated with low-nutrient soils, including coastal, montane, laterite and sandy, or areas subject to erosion or waterlogging.

GRASSLAND Generally dominated by herbaceous (non-woody) species and occurring in a range of areas. They can be split into two main types:

Savannah grassland is typical of tropical areas of Australia, with large areas of grasses and scattered trees. The climate, with extended droughts followed by

Heath

monsoon rains, is the key factor in the promotion of this grassland type. The soil is typically porous, allowing quick drainage of the vast amounts of water that fall during the 'wet' season. The 'dry' season can see numerous, sometimes large wildfires, both naturally occurring and lit by humans, which can burn through vast areas. The scattered trees typically retain enough moisture to survive the fires, but the parts of the grasses that are visible above ground are mostly destroyed. The arrival of the monsoon rains stimulates new growth from the underground rootstock and stored seeds.

Temperate grassland can be divided into **tussock grassland**, characterized by a broad range of perennial grasses growing in tufts (including Mitchell Grass and Blue Grass), and **hummock grassland**, which is dominated by spinifex (*Triodia* and *Plechrachne* spp.), and is typical of the arid lands of Australia. Evergreen perennials form mounds up to a metre in height, with areas of open, exposed and usually bare soil in between hummocks. Soils are typically sandy or rocky (skeletal), and either hilly or flat. This grassland type has been extensively cleared for grazing, often being replaced with exotic species, or modified by invasion of weed species or frequent burning.

MANGROVES These are found in the intertidal zone in coastal areas that are protected from high waves. Mangroves tend to form tall, closed forests in the north, and low, open forests or shrubland in the south. No native frog species rely exclusively on this habitat, although some species have been recorded on the fringes. The introduced Cane Toad has been recorded in mangroves.

INLAND WATERWAYS A mixture of fresh and brackish aquatic areas, including rivers, creeks, billabongs, lakes, swamps and marshes. Fresh water has an average salinity of less than 5g of salts per 1kg of water, while brackish water can range from 5 to 29g/kg. Australia also has numerous inland salt lakes, which can have salinity levels as high as 300g/kg. Freshwater areas are the primary habitats for most frog species, which require this essential resource for breeding purposes and development of aquatic larvae (tadpoles)

Peter Rowland

Paperbark Swamp

in species with this development stage. Some freshwater habitats are ephemeral, arising as a result of recent rains, and can disappear as quickly as they appear, while other waterways are more permanent in nature. Different frog species have adapted to the many types of freshwater habitat – some prefer flowing waterways, others are found around deep, permanent still-water ponds and lakes or shallow marshes, while others still have adapted to successfully utilize the fleeting ephemeral pools or damp soils and grasses.

Sand ridge desert

Peter Rowland

BARE GROUND Areas that are largely lacking in vegetation, or with some pioneer plant species. They can be in the form of exposed rock, coastal sands and dunes, desert sands and claypans. The soils have low nutrient content and are prone to erosion. Bare ground can exist in other habitats, such as woodland and grassland, associated with rivers and creeks, or it can form vast areas, becoming the dominant habitat type in a region.

URBAN AREAS These are diverse and include remnants of bushland areas (see above), but also contain wasteland, dwellings and gardens, which are homes to large numbers of invertebrate animals and thus provide a habitat for native frogs. Garden ponds and moist gardens can provide homes for many local frog species, although the high amount of pesticides, herbicides and other household chemicals used in many gardens makes many areas unsuitable for permanent frog populations.

FROG CALLS – AN AID TO IDENTIFICATION

Frog lungs are quite feeble, so frog calls are not particularly powerful, but the animals are able to amplify their calls using an inflatable throat pouch referred to as the vocal sac. The amount of amplification varies between species, with some having a call that can be audible over a kilometre, while others have calls that travel only a short distance. Some species use the environment to amplify their calls by calling into naturally occurring amphitheatres. Some species do not have an audible call at all, or live in noisy or dense environments where the call is quickly drowned out – these generally rely on hand or arm signals to attract a potential mate. Even when spotted, some species can closely resemble another that occupies the same geographic area. If this is the case, the call is almost essential for confirmation of identification. Most frogs, however, rely on camouflage or concealment for their protection, so are often cryptically coloured and patterned, or call from a concealed location, to avoid detection by potential predators, making them difficult to find.

Frog species emit a variety of calls, which can be given at any time of the year, particularly after rainfall, although the most commonly heard one is the mating call given by the male during the breeding season, when he tries to outsing his rivals to attract a female. At this time large numbers of males congregate around suitable breeding sites, producing an almost deafening noise. A male will attempt to select the best site to call from, so that he gives himself the best possible chance of attracting a mate. This can lead to wrestling bouts between males and specialized territoriality calls.

There are a number of devices that can assist in identifying a frog by its call. Over the years there have been several commercially available cassettes or CDs that contain recordings of various frog calls, and provide a sample of the call that can assist a user in matching up a call they hear in the wild with the one in the recording. There are now a few apps available that can be downloaded to a phone, tablet or other similar device. One of these, namely the FrogID app, developed by the Australian Museum (see www.frogid.net.au), records the frog call directly to your device, then matches it to similar calls held in the software's database.

FROG-FRIENDLY GARDENS

Frogs are popular and interesting animals to have in a garden. They can be indicators of a healthy environment, and also assist in reducing the abundance of many of the less popular insects, such as mosquitoes. Although the needs of frogs are basically straightforward (water, shelter and food), attracting frogs to your garden is not simply a matter of digging a hole and filling it with water. Frog-friendly gardens require some planning, and a local fauna group or frog society (see p. 171), can give you some expert advice specifically for your area. Do not capture frogs (or tadpoles), then release them to your garden, as this can introduce disease to the frogs that would naturally occur there. If you build a frog-friendly garden, they will come of their own accord. In general, the key steps to making your garden frog friendly include:

- The garden should be clean and free from potential contaminants. As mentioned earlier (see p. 6), frogs are sensitive to pesticides, herbicides, pollutants, detergents and other chemicals that are often used around the home and garden. They absorb these through their skin and from the food they eat, so eliminating non-organic chemicals from your garden will go a long way to preserving the frogs that do take up residence. Taking these measures also helps other native fauna in your garden.
- One of the key requirements for attracting frogs is the presence of moisture. This is vital in assisting to keep their skin moist, thus enabling them to dissolve oxygen through it, although they are also able to breathe through the nostrils and mouth. Moisture also attracts insect and other invertebrates, which form the bulk of their food. A well-vegetated garden (preferably with plants that are native to the area) will assist in retaining moisture in the soil and trapping humid air around frogs.
- Most frog species are active mainly at night. During the day, they often seek sheltered, cool and moist sites to hide from potential predators. They mostly hide among thick plant foliage, in moist ground debris (such as leaf litter), or under logs, rocks or plant pots. Providing such places will protect frogs from most diurnal predators. Cats are a major predator of wildlife in most of Australia, especially around urban and suburban gardens. If you own a cat, do not let it out at night when frogs are active.

- Frogs need food for vital metabolic processes, and actively hunt insects and other invertebrates. The more vegetated and moist your garden, the more insects it will attract. If you compost vegetable scraps and paper waste, this will also help to attract invertebrates that feed on this nutrient-rich organic matter, which in turn provides an ideal food supply for frogs. Insects are also attracted to artificial outdoor light sources (like solar lights), and a few of these strategically placed can attract frogs to the areas where the lights are placed.
- Finally, frogs need suitable sites in which to breed and lay their eggs, with a high ratio of shade to sunlight. For most species this will be a constant waterbody, such as a densely vegetated pond or water-filled cavity in an epiphytic plant or hollow. If installing a pond, ensure that it has varied depths, has sufficient aquatic plants to provide protection from predators and sunlight, and is able to be easily entered and exited by the frogs (frogs can drown if they become trapped in water). Increasing the density of vegetation around the edges of a pond will decrease the likelihood of it being used by the Cane Toad if it occurs in your area (also note that Cane Toads lay their eggs in long strands, which can be easily detected and removed if found). A frog-friendly bog, swamp or pond is ideally best sited in lower areas of a garden, where rainwater can naturally keep it topped up and the surrounding soil moist. Also keep in mind that if filling your pond from a tap that is connected to the water mains, chlorinated water should be allowed to stand for 4–5 days before it is added to the frog environment or use a water dechlorinator available from pet shops.

By researching the frogs that are in your area and supplying the type of environment they prefer, you can ensure that successive generations of these species will make your garden their home throughout the year. Always ensure, however, that your garden is a safe place, especially for young children, and keep in mind that frogs will call at night, so make sure that the best places for them to congregate are not right outside a bedroom window.

Peter Rowland

Frog pond in an urban garden

USING THIS BOOK

This guide provides up-to-date information on each of Australia's frogs. It is an introductory guide designed to help readers identify species. To help in identification, a photograph is included for each species described, though clear identification of many species relies on the use of call characteristics, genetics and sometimes larval morphology. The taxonomy follows Cogger 2018 with the exception of newly decribed taxa. There is a pending significant revision of the Australasian tree frogs that will change genus-level taxonomy for many of Australia's species (see p. 169 for table summarizing the changes).

DISTRIBUTION KEY & ABBREVIATIONS

Qld	Queensland		
NSW	New South Wales	**WA**	Western Australia
ACT	Australian Capital Territory	**NT**	Northern Territory
Vic	Victoria	**PNG**	Papua New Guinea
Tas	Tasmania	**GDR**	Great Dividing Range
SA	South Australia	**asl**	above sea level

KEY FEATURES & MEASUREMENTS

Sizes quoted in species accounts for body measurements are average maximum sizes (total length, abbreviated to TL), but exceptions can occur. Breeding information (such as clutch sizes, hatching periods and length of larval development) is taken from current literature and should be treated as an indicative value, as ongoing research can change the values provided.

PARTS OF FROGS

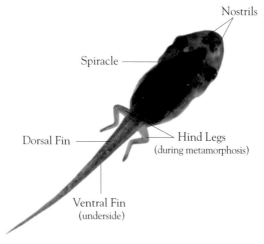

Nostrils

Spiracle

Dorsal Fin

Hind Legs
(during metamorphosis)

Ventral Fin
(underside)

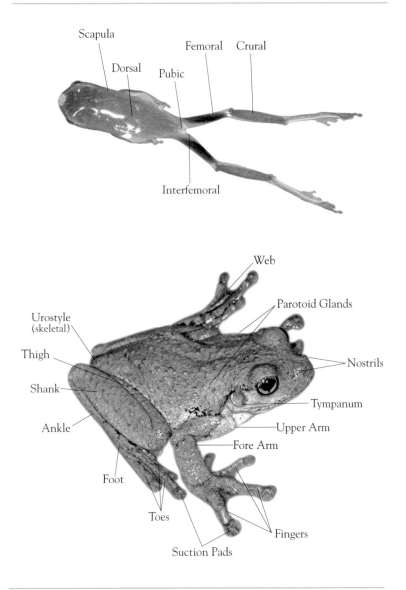

Scapula

Dorsal

Pubic

Femoral   Crural

Interfemoral

Web

Parotoid Glands

Urostyle
(skeletal)

Thigh

Shank

Ankle

Foot

Nostrils

Tympanum

Upper Arm

Fore Arm

Toes

Fingers

Suction Pads

Glossary

amplexus Sexual embrace or mating embrace in frogs.

aquatic Habitat in or near water.

arboreal Individuals that live predominantly above the ground, for example in vegetation.

arthropod Animal with an external skeleton. Predominantly refers to insects but also includes spiders and crustaceans.

basking Act of a frog exposing itself to increased temperature in order to raise its core body temperature.

brumate/brumation Behaviour akin to mammalian hibernation in which a frog seeks out a secluded spot, ceases feeding and lowers its metabolic rate to wait out the cold winter period.

canthal Referring to a ridge that runs between the snout and the eyes.

cathemeral Active at any time of day or night.

caudal At or near rear half of body.

chitin Major component of hard outer coating or exoskeleton of an arthropod.

cloaca Combined exit for gastrointestinal waste, kidney waste and reproductive products.

clutch Number of eggs laid by a female frog in a single reproductive event.

colour morph Any colour that is different from normal colour.

crepuscular Active at dawn and dusk.

cryptic Disguised appearance, either through colour and pattern or habits.

diurnal Active during daytime.

dorsal Of, on, or relating to upper half or top of a structure or body.

ephemeral pool Temporary or semi-permanent pool, adjacent to another watercourse where the water is still. Such pools are filled when the watercourse floods.

epiphytic Refers to plant species that grow on surfaces of other plants without deriving nutrition from those plants (like an orchid).

extralimital Referring to population occurring outside the primary accepted habitat range – for example, a typically PNG species that may be found in Australia.

fossorial Living or active beneath soil surface.

gene Basic unit of genetic control. Each gene has a specific function and is found on a specific section of a specific chromosome.

gravid Pregnant; the abdominal cavity contains formed eggs or young.

heliothermic Needing to bask to raise body temperature.

herbivorous Feeding predominantly on plant material.

herpetofauna Collective term referring to group of amphibians and reptiles – usually pertaining to particular region, locality or habitat.

hybrid Genetic combination as a result of mating of two different species or subspecies under unnatural circumstances, as may occur in captivity. Some people also consider mating of individuals from different localities to be a form of hybridization.

inguinal Of the groin area.

insectivorous Feeding predominantly on insects.

intergrade Long-term genetic combination of two different species or subspecies as a result of pairing under natural circumstances, such as at the natural boundary between two overlapping distributions, resulting in a population displaying variable features between the two species or subspecies.

iris Part of eye surrounding pupil.

juvenile Young individual that cannot yet be visually sexed and still possesses some colouration of a hatchling.

metamorphosis Act in which an animal changes physical state from tadpole to frog.

metatarsal tubercle A fleshy protrusion on the posterior edge of the rear feet that aids in digging.

morphological Pertaining to external appearance of an animal.

mutation A mutant gene is created, resulting in change in the genetic make-up of an individual – usually seen as a change in physical or body colouration.

nocturnal Active during the night.

nuchal Area where head and neck join.

nuptial pads Dark-coloured pads on front feet of male frogs that enlarge when frog is reproductively stimulated. The pads are quite rough and aid in amplexus.

parotoid Referring to a gland under the skin, just behind eye, which can contain toxins used in defence.

seepage (soak) Water slowly leaking from underground or adjacent area through porous material or holes.

spawn Clutch of frog eggs. The act of spawning is when the female releases her eggs into the water as the male sprays sperm over them, resulting in them being fertilized.

stippling Patterns or markings created by grouping together of numerous small dots.

subspecies Taxonomic category that is a variation in a primary (nominate) species brought about by geographical or genetic isolation, usually characterized by a variation in morphological features.

sympatry Individuals that share the same habitat.

taxonomy Study of genetic and morphological relationships between species.

terrestrial Individuals that live on the ground surface.

tibial gland A gland found on the rear leg between knee and ankle.

tubercle Small, rounded nodule or projection.

tympanum Membrane covering entrance to ear of a frog.

ventral/ventrum Relating to lower half of bottom of a structure or body.

> **LIMNODYNASTIDAE (SWAMP FROGS)**
> This Australasian family of terrestrial frogs occurs across Australia and into New Guinea.
> It comprises foam nest breeders that differ significantly in morphology and ecology.

Tusked Frog ▪ *Adelotus brevis* TL 51mm

DESCRIPTION Upperparts grey to brown or blackish, mottled with darker pigment forming spots and flecks, and with low, rounded tubercles. Underside black to grey with white markings. Bright red flashing colours with darker banding in groin and armpit

extending on to flanks and belly. Two prominent 'tusks' inside jaw of males, which are modified teeth, used in male combat. **DISTRIBUTION** Found in eastern Qld, from Eungella to Ourimbah, NSW. **HABITS AND HABITAT** Nocturnal. Occurs in rainforests and wet forests near still pools, where it feeds on invertebrates. Male's call is a *chuck* similar to that of a chicken. Lays about 340 eggs that hatch into tadpoles, that become frogs after about 7 weeks.

Western Spotted Frog ▪ *Heleioporus albopunctatus* TL 90mm

DESCRIPTION Upperparts dark brown, becoming lighter on flanks, and regularly spotted with white to yellowish-cream. Skin surface granular. Underparts creamy-white. Large black spine on inside of first finger on males that is used during amplexus. **DISTRIBUTION** Found in southwestern WA, from just west of Esperance, north to Kalbarri. **HABITS AND HABITAT** Nocturnal. Inhabits mulga, mallee, open woodland, swamps and claypans, where it lives in burrows in soft soil. Feeds on invertebrates. Call is a short *whoop* repeated about a second apart. About 500 eggs are laid in a burrow, which hatch when burrow fills with water. Tadpoles metamorphose after around 24 weeks.

Giant Burrowing Frog ■ *Heleioporus australiacus australiacus* TL 80mm
(Eastern Owl Frog)

DESCRIPTION Grey to dark purple above with lighter greyish marking on flanks and fine yellow to orange spotting along sides. Underparts whitish. Back has numerous small bumps, giving a smooth but granular texture. **DISTRIBUTION** Ranges from Watagan Mountains to Dharwal NP in eastern NSW. Range has contracted significantly, and is no longer present at many historic sites in north-east of range. **HABITS AND HABITAT** Nocturnal. Lives in open woodland, closed woodland, heaths, swamps and streams, where it feeds on

invertebrates. Call, a frequently repeated *oot oot oot*, similar to that of an owl, is emitted in a series from the edge of a waterbody, or occasionally from burrows constructed in banks. The call has a lower pulse rate than the closely related Southern Owl Frog (see below). About 1,200 eggs, protected by a foam nest, are laid in a burrow or under thick vegetation that is constructed by both male and female. Tadpoles start to become metamorphlings after about 12–46 weeks, depending on pool temperature and food availability.

Scott Eipper

Southern Owl Frog ■ *Heleioporus australiacus flavopunctatus* TL 90mm

DESCRIPTION Grey to dark purple above with lighter greyish markings on flanks. Large yellow to orange spotting along sides, usually forming ring around cloaca. Underparts whitish. Back has numerous small bumps, giving a smooth but granular texture. **DISTRIBUTION** Parma Creek Nature Reserve, Kangaroo Valley, NSW, to Walhalla, eastern VIC. Range has contracted significantly; no longer present at many historic sites. **HABITS AND HABITAT** Nocturnal. Lives in open and closed woodland, heaths, swamps and streams, where it feeds on invertebrates. Call a frequently repeated *oot oot oot*, similar to that of an owl, emitted in series from edge of a waterbody or, occasionally, from burrows constructed in banks. Call has a faster pulse rate than that of closely related Giant Burrowing Frog (see above). About 900 eggs, protected by a foam nest, laid in burrow or under thick vegetation, constructed by male and female. Tadpoles start to become metamorphlings after about 13–26 weeks, depending on pool temperature and food availability.

Nick Clemann

Juvenile

Michael Swan

Hooting Frog ■ *Heleioporus barycragus* TL 86mm

DESCRIPTION Grey to dark purple above with lighter greyish markings on flanks and yellow to orange spotting along sides. Back has small bumps but is mainly smooth. Skin surface granular. Underside creamy-white. Large black spine on inside of first finger

on males that is used during amplexus. DISTRIBUTION Found in southwestern WA, from Darling Range to Dryandra Woodland. HABITS AND HABITAT Nocturnal. Lives in rocky closed woodland, open woodland, swamps and claypans, where it inhabits burrows in clay loamy soil. Feeds on invertebrates. Call is a short *whoop* repeated about a second apart. About 500 eggs are laid in a burrow, which hatch when burrow fills with water. Tadpoles metamorphose after about 29 weeks.

Angus McNab

Moaning Frog ■ *Heleioporus eyrei* TL 65mm

DESCRIPTION Upperparts grey to brown above with lighter greyish marking on flanks, and usually yellowish above shoulders. Most individuals mottled and flecked with both lighter and darker markings. Skin surface granular. Underside creamy-white. Large black spine on inside of first finger on males that is used during amplexus. DISTRIBUTION

Found in southwestern WA, mainly along coastal fringes from Geraldton to Cape Arid. HABITS AND HABITAT Nocturnal. Inhabits grassland, open woodland, swamps and claypans, where it lives in burrows in soft sandy soil. Feeds on invertebrates. Call is a rising moan repeated about 2 seconds apart. About 300 eggs are laid in a burrow, which hatch when burrow fills with water. Tadpoles are thought to metamorphose after 16–25 weeks.

Scott Eipper

Plain Frog ■ *Heleioporus inornatus* TL 75mm

DESCRIPTION Upperparts purplish-grey to brown. Some individuals immaculate, others weakly mottled and flecked. Underside creamy-white. Skin surface granular. One or two small black spines on inside of first finger on males, used during amplexus.

DISTRIBUTION Found in southwestern WA, from Darling Range to Albany. **HABITS AND HABITAT** Nocturnal. Inhabits grassland, open woodland and swamps, where it lives in burrows in soft sandy or peaty soils. Feeds on invertebrates. Call is a rapidly repeated *whoop* that is similar to sounds in a video game. About 300 eggs are laid in a burrow, which hatch when burrow fills with water. Tadpoles metamorphose after about 16 weeks.

Scott Eipper

Sand Frog ■ *Heleioporus psammophilus* TL 63mm

DESCRIPTION Upperparts purplish-grey to brown, usually heavily mottled but sometimes immaculate. Underside creamy-white. Skin surface granular. One small black spine on inside of first finger of male, used during amplexus. **DISTRIBUTION** Found in southwestern WA, from Geraldton to Albany. **HABITS AND HABITAT** Nocturnal.

Lives in grassland, open woodland and swamps, where it occupies burrows in soft sandy soils. Feeds on invertebrates. Call is short and cricket-like, and is the only reliable trait that distinguishes this species from the Plain Frog (see above) in the field. About 250 eggs are laid in a burrow, which hatch when burrow fills with water. Duration as tadpoles unknown.

Scott Eipper

Sandpaper Frog ▪ *Lechriodus fletcheri* TL 59mm

DESCRIPTION Colouration tan or light brown to dark brown above, with darker bands on legs, and black band from back of eye, extending over tympanum to shoulder. Back rough, like sandpaper. Underside white, with dark grey pigment to hands and feet.

Some herpetologists place this species in the genus *Playplectrum*. **DISTRIBUTION** Found from Mt Tamborine, Qld, to Ourimbah, NSW. **HABITS AND HABITAT** Nocturnal. Lives around watercourses in rainforests, where it feeds on invertebrates. Male's call, a bobbling *roc roc roc roc*, is made while floating in water. Breeds in still rainforest pools, and about 450 eggs are laid in a foam nest. Tadpoles start to develop into metamorphlings after 4 weeks.

Marbled Frog ▪ *Limnodynastes convexiusculus* TL 60mm

DESCRIPTION Light brown to dark brown above, with darker olive, grey or dark green spotting and blotches. Usually a black band from back of eye, extending over tympanum to shoulder. Body covered in raised bumps. Underside white. **DISTRIBUTION** Found

from Townsville, Qld, across northern Australia to Kimberley, WA. Also in southern PNG. **HABITS AND HABITAT** Nocturnal. Lives around watercourses in savannah and swamps, where it feeds on invertebrates. Male's call, a resonating *honk*, is made while floating in water. Breeds in still waterbodies, and about 2,000 eggs are laid in a foam nest. Tadpoles start to develop into metamorphlings after 8 weeks.

Flat-headed Frog ■ *Limnodynastes depressus* TL 55mm

DESCRIPTION Light brown to dark brown above, with darker olive, grey or dark green spotting and blotches. Head greatly flattened with eyes high on top of head. Underside white. **DISTRIBUTION** Found from the Kimberley, WA, to north-west NT. **HABITS AND HABITAT** Nocturnal. Lives around watercourses in savannah, swamps and agricultural areas, where it feeds on invertebrates. For many years it was confused with the Spotted Marsh Frog (see p. 34). Male's call, similar to a ratcheting clicking, is made while floating in water. Breeds in still waterbodies, and about 2,000 eggs are laid in a foam nest. Tadpoles thought to start to develop into metamorphlings after 8 weeks.

Angus McNab

Western Banjo Frog ■ *Limnodynastes dorsalis* TL 75mm

DESCRIPTION Colouration and pattern variable. Yellow to light grey to almost black with darker spots, and with thin yellow to orange vertebral stripe on most individuals. Others are grey to chocolate-brown, with yellow and orange spotting along sides. Back mainly smooth with small bumps, and groin has bright red markings. Underside white. **DISTRIBUTION** Found in south-west WA, from Murchison River to Wattle Camp. **HABITS AND HABITAT** Nocturnal. Lives around watercourses in grassland, rocky gorges, swamps and agricultural areas, where it feeds on invertebrates. Male's call similar to a loud *bonk*, and made while floating in water. Breeds in still waterbodies, and around 2,000 eggs are laid in a foam nest. Duration as tadpoles unknown.

Angus McNab

Eastern Banjo Frog ▪ *Limnodynastes dumerilii dumerilii* TL 75mm

DESCRIPTION Underparts light grey to almost black with orange to yellow markings on lower flanks, and thin yellow to orange vertebral stripe on most individuals. Back covered with small bumps and tubercles. Underside yellow to orange with dark grey flecking. **DISTRIBUTION** Found from south-east Qld, through NSW, Vic and into south-east SA. **HABITS AND HABITAT** Nocturnal. Lives around slow-moving or still watercourses,

including artificial dams, flooded grassland, open woodland and roadside ditches, where it feeds on invertebrates. Male's call, an explosive *bonk*, repeated frequently, is made from beneath emergent vegetation along the edge of a waterbody. Eggs are laid and protected in a foam nest constructed by both male and female, with around 2,300 in a clutch. Tadpoles develop into frogs after about 23 weeks.

Snowy Mountains Banjo Frog ▪ *Limnodynastes dumerilii fryi* TL 83mm

DESCRIPTION Colouration light grey to almost black, with orange to yellow markings on lower flanks, and thin yellow to orange vertebral stripe on most individuals. Back covered with small bumps and tubercles. **DISTRIBUTION** Found in Snowy Mountains, NSW. **HABITS AND HABITAT** Nocturnal. Lives around slow-moving or still watercourses,

flooded grassland, open woodland and roadside ditches, where it feeds on invertebrates. Male's call, a frequently repeated, explosive *bonk*, is made from beneath emergent vegetation along the edge of a waterbody. Reproductive biology largely unknown, but expected to be similar to that of the Eastern Banjo Frog (see above), although tadpoles would most likely take longer to complete metamorphosis due to cooler temperatures.

Coastal Banjo Frog ■ *Limnodynastes dumerilii grayi* TL 65mm

DESCRIPTION Upperparts typically light grey to almost black, with orange to yellow markings on lower flanks, and thin yellow to orange vertebral stripe. Some individuals grey to chocolate-brown, with yellow and orange spotting along sides. Back covered with small bumps and tubercles. Underside yellowish with dark grey flecking. **DISTRIBUTION** Found from Nambucca to Jervis Bay, NSW. **HABITS AND HABITAT** Nocturnal. Lives

around slow-moving or still watercourses, including artificial dams, flooded grassland, open woodland and roadside ditches, where it feeds on invertebrates. Male's call, a frequently repeated, explosive *bonk*, is made from beneath emergent vegetation along the edge of a waterbody. About 2,000 eggs are laid in a clutch, protected in a foam nest that is constructed by both male and female. Tadpoles develop into frogs after about 32 weeks.

Scott Eipper

Southern Banjo Frog ■ *Limnodynastes dumerilii insularis* TL 65mm

DESCRIPTION Upperparts light bluish-grey to almost black, with orange to yellow markings on lower flanks, and thin yellow to white vertebral stripe on most individuals. Body usually spotted with olive to grey. Back covered with small bumps and tubercles. Underparts white to cream with light grey to pale brown flecking. **DISTRIBUTION** Found in southern Vic and Tas; possibly localized population at Jervis Bay, NSW. **HABITS AND HABITAT** Nocturnal. Lives around slow-moving or still watercourses, including artificial

dams, flooded grassland, open woodland and roadside ditches, where it feeds on invertebrates. Male's call, a frequently repeated *bonk*, is made from beneath emergent vegetation along the edge of a waterbody. About 3,500 eggs are laid in a clutch, which are protected in a foam nest constructed by both male and female. Tadpoles develop into frogs after about 23–62 weeks.

Scott Eipper

Mottled Banjo Frog ■ *Limnodynastes dumerilii variegatus* TL 65mm

DESCRIPTION Light bluish-grey to almost black above, with orange to yellow markings on lower flanks, and without a vertebral stripe. Body usually spotted with olive to grey. Back covered with small bumps and tubercles. Underside white to cream with dark grey mottling. **DISTRIBUTION** Found in south-west Vic, neighbouring SA and King

Island, Tas. **HABITS AND HABITAT** Nocturnal. Lives around slow-moving or still watercourses, including artificial dams, flooded grassland, open woodland and roadside ditches, where it feeds on invertebrates. Male's call, a frequently repeated *bonk*, is made from beneath emergent vegetation along the edge of a waterbody. Reproductive biology thought to be similar to that of the Coastal Banjo Frog (see p. 29).

Barking Marsh Frog ■ *Limnodynastes fletcheri* TL 58mm

DESCRIPTION Colouration and pattern vary from olive, to light grey, to charcoal with darker spots. Usually reddish markings over backs of eyes. Back smooth, without bumps or protrusions. **DISTRIBUTION** Found in Qld south of Mackay, NSW, northern Vic west of Great Dividing Range, across into SA via Murray River. **HABITS AND HABITAT** Lives

around slow-moving or still watercourses, including artificial dams, flooded grassland, open woodland and roadside ditches, where it feeds on invertebrates. Male's call, a single *wrrark*, is made from beneath emergent vegetation along edges of the water. Up to 1,500 eggs are laid and protected in a foam nest constructed by both male and female. Tadpoles develop into metamorphlings in about 10 weeks.

Giant Banjo Frog ■ *Limnodynastes interioris* TL 90mm

DESCRIPTION Upperparts grey to chocolate-brown with yellow and orange spotting along sides, and usually with orange tibial gland. Back covered with tubercles and small bumps. Underside yellow to orange, usually without spotting or flecking. **DISTRIBUTION**

Found from border region of Vic and NSW, west of Great Dividing Range north to Brewarrina. **HABITS AND HABITAT** Nocturnal. Lives around watercourses in grassland, dry woodland, swamps and agricultural areas, where it feeds on invertebrates. Male's call resembles a loud, resonating *bonk*, and is made while floating in water. Breeds in still waterbodies, and about 1,000 eggs are laid in a foam nest. Tadpoles start to develop into metamorphlings after 25 weeks.

Scott Eipper

Carpenter Frog ■ *Limnodynastes lignarius* TL 63mm

DESCRIPTION Colouration light brown to grey above, with darker spotting and blotches. Blotches are black to charcoal to dark brown in east of range, but typically reddish in western population. Body greatly flattened, probably as an adaptation to living in rock crevices, and tympanum large and distinct. Underparts white. **DISTRIBUTION** Found from Kimberleys, WA, to northern NT. **HABITS AND HABITAT** Nocturnal. Lives

Scott Eipper

around watercourses in escarpment country and gorges, where it shelters in moist rock crevices. Feeds on invertebrates. Possibly two species, with western individuals having different calls and colouration. Call sounds like pieces of timber being hit together and has a higher tone in western individuals. Breeds in still waterbodies such as rock pools, and about 350 eggs are laid in a foam nest. Tadpoles start to develop after 9 weeks.

Western form

Adam Elliott

Striped Marsh Frog ■ *Limnodynastes peronii* TL 75mm

DESCRIPTION Colouration and pattern variable, from yellow to light grey to almost black, with darker and lighter stripes running along body. Back smooth without bumps or protrusions. **DISTRIBUTION** Found in eastern Australia, from north-east Qld to south-east SA, including northern Tas. **HABITS AND HABITAT** Lives around slow-moving or still watercourses, including man-made dams, flooded grassland, lakes and ornamental pools, as well as along streams and roadside ditches. Call sounds like *tok* and is repeated

frequently, and given from beneath emergent vegetation along edges of a waterbody. Call can be heard year round, but is most frequently given after rain. Typically breeds in permanent pools, but will also breed in ephemeral pools alongside waterways. Up to 1,000 eggs are laid and protected by a foam nest. Tadpoles change into frogs after about 6 weeks.

Salmon-striped Marsh Frog ■ *Limnodynastes salmini* TL 73mm

DESCRIPTION Light grey to charcoal with darker blotches, and usually with pair of yellow to orange-red stripes extending from shoulder to groin. Some individuals have a yellow triangle between snout and eyes. Usually a broken-edged dark stripe from snout. Back smooth, without bumps or protrusions. **DISTRIBUTION** Found in Qld, south of Mackay, and through most of central NSW to Wagga Wagga. **HABITS AND HABITAT**

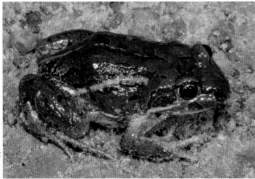

Lives around slow-moving or still watercourses, including artificial dams, flooded grassland, open woodland and roadside ditches. Male's call, a single, musical *wrrark*, is made from beneath emergent vegetation along edges of water. Up to 2,000 eggs are laid in a clutch, protected in a foam nest. Tadpoles develop into frogs after about 6 weeks.

Spotted Marsh Frog ■ *Limnodynastes tasmaniensis* TL 49mm

DESCRIPTION Colouration and pattern vary from yellow, to light grey, to almost black, with darker spots. Thin orange to red vertebral stripe on most individuals. Back smooth, without bumps or protrusions. **DISTRIBUTION** Found over much of Qld, NSW, Vic, Tas and into eastern SA. **HABITS AND HABITAT** Lives around slow-moving or still watercourses, including artificial dams, flooded grassland, lakes

and ornamental pools, as well as beside streams and roadside ditches. Male's call, sounding like *tik tik tik*, is repeated frequently, and given from beneath emergent vegetation along edges of water. Calls most frequently after rain, but can be heard year round. Breeds in ponds or pools, preferring little or no surface movement. Up to 1,350 eggs are laid in a clutch, protected in a foam nest. Tadpoles develop into metamorphlings in 13–20 weeks.

Northern Banjo Frog ■ *Limnodynastes terraereginae* TL 79mm

DESCRIPTION Colouration and pattern variable. Typically yellow to light grey to almost black, with darker spots, but some individuals are grey to chocolate-brown with yellow and orange spotting along sides. Back mainly smooth with small bumps, and groin has bright red markings. Underside white, with red flecking beneath rear limbs. **DISTRIBUTION** Found from north-east Qld to northeastern NSW. **HABITS AND HABITAT** Nocturnal. Lives around watercourses in grassland, dry woodland, swamps and agricultural areas, where it feeds on invertebrates. Male's call similar to a loud *bonk*, and made while floating in water. Breeds in still waterbodies, with about 2,000 eggs laid in a foam nest. Tadpoles start to develop into frogs after 10 weeks.

White-footed Frog ■ *Neobatrachus albipes* TL 46mm

DESCRIPTION Grey to brown above, usually with black mottling, lighter on lower flanks. Tops of feet white, and metatarsal tubercle pale brown or white. Body smooth, without tubercles. Underside cream to white without markings. **DISTRIBUTION** Restricted to

southern WA, from Cape Arid, west to Albany and north to Coolgardie. **HABITS AND HABITAT** Nocturnal. Lives in swamps, bushland and surrounding agricultural areas. Likely to live below the ground in self-made cocoons to prevent water loss, emerging at night, usually after rain, to feed on invertebrates. Male's call, a low trill, is made from beneath vegetation. About 1,000 eggs are laid in still pools. Length of time as tadpoles usually about 10 weeks, but can be much longer if they develop over winter.

Northern Trilling Frog ■ *Neobatrachus aquilonius* TL 59mm

DESCRIPTION Reddish-brown to grey above, with extensive yellow mottling, lighter on lower flanks. Body smooth, without tubercles. Underside cream to white without

markings. Metatarsal tubercle clear in colour. **DISTRIBUTION** Found over most of half of western NT and much of inland WA. **HABITS AND HABITAT** Nocturnal. Inhabits claypans, deserts and open woodland, where it is likely to live below the ground in self-made cocoons to prevent water loss, emerging at night, usually after rain, to feed on invertebrates. Male's call is a low trill, made from beneath vegetation. About 1,400 eggs are laid in still pools. Length of time as tadpoles unknown.

Tawny Trilling Frog ■ *Neobatrachus fulvus* TL 50mm

DESCRIPTION Upperparts yellowish-brown to grey with extensive mottling, paler yellow on lower flanks. Body smooth, without tubercles. Underside cream to white without markings. Metatarsal tube has black margins. **DISTRIBUTION** Found in coastal WA, from Shark Bay to North West Cape. **HABITS AND HABITAT** Nocturnal. Lives in claypans, deserts and open grassland. Likely to live below the ground in self-made cocoons to prevent water loss, emerging at night, usually after rain, to feed on invertebrates. Male's call, a high-pitched trill, is made from beneath vegetation. Reproductive biology thought to be similar to that of the Kunapalari Frog (see below).

Brad Maryan

Kunapalari Frog ■ *Neobatrachus kunapalari* TL 59mm

DESCRIPTION Highly variable. Yellowish-brown to grey with extensive mottling. Body smooth, without tubercles in females, while males have short black tubercles across front half that enlarge into spines during breeding season. Underside cream to white without markings. Metatarsal tubercle has black margins. **DISTRIBUTION** Found in southern WA, from edge of Nullarbor Plain, through Goldfields region to about Kalbarri. **HABITS AND HABITAT** Nocturnal. Lives in claypans, mallee, deserts and open grassland. Likely to live below the ground in self-made cocoons to prevent water loss, emerging at night, usually after rain, to feed on invertebrates. Male's call, a high-pitched trill, is made from beneath vegetation. About 750 eggs are laid in still pools, and after hatching remain as tadpoles for about 20 weeks.

Adam Elliott

Humming Frog ■ *Neobatrachus pelobatoides* TL 45mm

DESCRIPTION Highly variable, from greenish-yellow to grey, reddish, tan or brown, often extensively mottled, and usually with thin red or white line between eyes and down midline. Body has many short tubercles. Underparts cream to white without markings. Metatarsal tubercle has black margins. **DISTRIBUTION** Found in southern WA, from edge of Israelite Bay, through goldfields region, reaching coast just north of Perth, and north to about Tamala.

Grant Webster

HABITS AND HABITAT Nocturnal. Inhabits claypans, mallee, deserts and open grassland. Likely to live below the ground in self-made cocoons to prevent water loss, emerging at night, usually after rain, to feed on invertebrates. Male's call is a low hum, and is made from beneath vegetation. About 530 eggs are laid in still pools, and after hatching remain as tadpoles for about 20 weeks.

Painted Trilling Frog ■ *Neobatrachus pictus* TL 55mm

DESCRIPTION Highly variable. Typically yellowish-brown to grey above with extensive mottling. Body has small tubercles, and skin of groin is loose. Underside cream to white without markings. Metatarsal tubercle black. **DISTRIBUTION** Occurs from Eyre Peninsula, southern SA, to Wedderburn, Vic.

Scott Eipper

HABITS AND HABITAT Nocturnal. Inhabits claypans, mallee, deserts and open grassland, where it lives below the ground in self-made cocoons to prevent water loss. Emerges at night, usually after rain, and feeds on invertebrates. Male's call, a high-pitched trill, is made from beneath vegetation. About 750 eggs are laid in still pools, and after hatching remain as tadpoles for about 25 weeks.

Sudell's Trilling Frog ■ *Neobatrachus sudellae* TL 55mm

DESCRIPTION Highly variable. Typically yellowish-brown to grey with extensive mottling, often with pale vertebral stripe. Body covered in small tubercles. Underside cream to white without markings. Metatarsal tubercle has black margins. **DISTRIBUTION** Found in southern Australia, from about Mt Magnet, WA, through southern NT and SA, to outskirts of Brisbane, Qld, and extending south to Sydney, NSW, and Melbourne, Vic.

HABITS AND HABITAT
Nocturnal. Inhabits claypans, grassland, mallee and open woodland, where it lives below the ground in self-made cocoons to prevent water loss. The species has been revised and now includes what was previously *N. centralis*. Emerges at night, usually after rain, to feed on invertebrates. Male's call is a high-pitched trill, made from beneath vegetation. Around 850 eggs are laid in still pools, and after hatching the tadpoles take about 25 weeks to metamorphose.

Scott Eipper

Shoemaker Frog ■ *Neobatrachus sutor* TL 51mm

DESCRIPTION Colouration yellow to gold with brown to black spotting. Body smooth, generally without tubercles. Underside cream to white without markings. Metatarsal tubercle has black margins. **DISTRIBUTION** Found over much of WA across into south-west NT. **HABITS AND HABITAT**
Nocturnal. Lives in claypans, mallee, deserts and open grassland. Likely to live below the ground in self-made cocoons to prevent water loss, emerging at night, usually after rain, to feed on invertebrates. Male's call, which resembles the sound of a small hammer tapping on timber, is made from beneath vegetation. About 1,000 eggs are laid in still pools, and tadpoles take about 4 weeks to metamorphose.

Scott Eipper

Plonking Frog ■ *Neobatrachus wilsmorei* TL 6cm
(Willsmore's Trilling Frog)

DESCRIPTION Colouration brown to grey-black with yellow stripes, and often with yellow vertebral stripe. On flanks yellow stripes often form yellow-edged dark triangle.

DISTRIBUTION Occurs in central WA, from north of Shark Bay through Goldfields region to about Lake Nabberu. **HABITS AND HABITAT** Nocturnal. Lives in claypans, deserts and open grassland. Likely to live below the ground in self-made cocoons to prevent water loss, emerging at night, usually after rain, to feed on invertebrates. Male's call, a single, hollow-sounding click, is made from beneath vegetation. Length of time as tadpoles is about 6 weeks.

Crucifix Spadefoot ■ *Notaden bennettii* TL 70mm
(Crucifix Toad)

DESCRIPTION Colouration yellow to olive. Body capped in black, brown, red and dark green blunt tubercles, which usually form cross-shaped marking over back. Underside cream to white without markings. Body rounded in shape. Metatarsal tubercle pale in colour. **DISTRIBUTION** Found west of Great Dividing Range, NSW, and southern Qld.

HABITS AND HABITAT Nocturnal. Lives in dry woodland, deserts, black-soil plains and brigalow. Lives in burrows deep below the ground, emerging at night, usually after heavy rain. and feeding on ants and termites. Male's call, a short *hoot* repeated every second or so, is made while floating in water. Around 1,000 eggs are laid in still pools, and after hatching tadpoles usually take about 7 weeks to metamorphose.

Northern Spadefoot ■ *Notaden melanoscaphus* TL 60mm

DESCRIPTION Grey to brown above, often with four blotches that join to form red and yellow cross, much brighter in younger adults and juveniles but typically with minimal pattern as older adults. Body capped in black, brown, red and mustard blunt tubercles. Underside cream to white without markings. Body rounded in shape. Metatarsal tubercle black in colour. **DISTRIBUTION** Found across northern Australia, from Georgetown, Qld, to Kimberley, WA. Isolated population near Bluewater, Qld. **HABITS AND HABITAT** Nocturnal. Inhabits dry woodland, deserts, black-soil plains and savannah, where it lives in burrows deep below the ground, emerging at night, usually after heavy rain, to feeds on ants and termites. Male's call, a short *hoot* repeated every second or so, is made while floating in still water. About 830 eggs are laid in still pools, and tadpole stage usually lasts about 8 weeks.

Adam Elliott

Desert Spadefoot ■ *Notaden nichollsi* TL 67mm

DESCRIPTION Upperparts yellow to olive. Body rounded and capped in black, brown, red and dark green blunt tubercles, which usually form indistinct darkened cross-shaped marking over back. Underside cream to white without markings. Metatarsal tubercle pale in colour. **DISTRIBUTION** Found in western Qld from Thargomindah, through northern SA and southern NT, north to Tennant Creek, across to Derby, WA. **HABITS AND HABITAT** Nocturnal. Lives in dry woodland, deserts and claypans, where it inhabits burrows deep below the ground, emerging at night, usually after heavy rain, to feed on ants and termites. Male's call, a short *hoot* repeated every second or so, is made while floating in still water. About 1,000 eggs are laid in still pools, which after hatching remain as tadpoles for around 3–4 weeks.

Scott Eipper

Weigel's Spadefoot ■ *Notaden weigeli* TL 80mm

DESCRIPTION Orange above with grey flanks and red spotting on legs. Body rounded and capped in red and brown blunt tubercles. Cream to white below without markings. Metatarsal tubercle pale in colour. **DISTRIBUTION** Restricted to Mitchell Plateau and

neighboring areas in WA. **HABITS AND HABITAT** Nocturnal. Inhabits dry woodland associated with sandstone gorges and spinifex, where it lives in deep rock crevices and burrows, emerging at night, usually after rain, to feed on ants and termites. Male's call, a short *hoot* repeated every second or so, is made while floating in still water. Reproductive biology largely unknown.

Baw Baw Frog ■ *Philoria frosti* TL 55mm

DESCRIPTION Grey or dark brown to blackish above, sometimes with yellow mottling. Back smooth with a few low, scattered tubercles. Prominent parotoid glands. Cream below with or without brown flecks and blotches. **DISTRIBUTION** Now found only in a couple of locations on Mt Baw Baw, Vic, above 1,100m asl. Historically found over much of Baw Baw National Park. **HABITS AND HABITAT** Cathemeral. Lives in seepages in montane,

wet sclerophyll forests, where it feeds on invertebrates. Has suffered massive declines largely due to chytrid fungus, and likely to become extinct without significant intervention. Researchers have begun setting up insurance populations. Male's call, a guttural *wrocc* repeated in short bursts, is made from a burrow. About 100 eggs are laid in a foam nest, and tadpoles start to develop into metamorphlings after 7 weeks.

Mount Ballow Mountain Frog ■ *Philoria knowlesi* TL 32mm

DESCRIPTION Reddish-brown, bronze or pale tan above; occasionally dark greenish-brown, usually with dark stripe from snout, through eye, to above shoulder; often with white upper edge. Body smooth. Reddish-yellow to dark brown below with pale flecking. **DISTRIBUTION** Found in a few upland locations on south-east QLD and NSW border, between Lever's Plateau and Mount Barney National Park. **HABITS AND HABITAT** Cathemeral. Lives in seepages in rainforests, where it feeds on invertebrates. Male's call, a guttural *wrocc*, made from within chamber of mud, usually beneath rocks and leaf litter. About 100 eggs laid in foam nest, and tadpoles start to develop into metamorphlings after seven weeks.

Jono Hooper

Red and Yellow Mountain Frog ■ *Philoria kundagungan* TL 32mm

DESCRIPTION Reddish-orange to dark brown or yellow above, usually with dark stripe from rear of eye to above shoulder. Body smooth. Underside bright yellow, occasionally with bright red flash markings. **DISTRIBUTION** Found in a few upland locations on south-east Qld and NSW border. **HABITS AND HABITAT** Cathemeral. Lives in seepages in rainforests, where it feeds on invertebrates. Male's call, a deep, repeated *wrocc*, is made from within a chamber of mud, usually beneath rocks and leaf litter. Little is known about its reproductive habits, but they are thought to be similar to those of the Sphagnum Frog (see p. 43). Females have been seen with active nests containing tadpoles.

Scott Eipper

Loveridge's Frog ■ *Philoria loveridgei* TL 33mm

DESCRIPTION Light grey, tan to dark brown above, typically with dark stripe from snout, through eye to above shoulder, often with white upper edge. Body smooth. Cream to dirty yellow below with darker flecking. **DISTRIBUTION** Found in a few upland locations on

south-east Qld and NSW border. **HABITS AND HABITAT** Cathemeral. Lives in seepages in rainforests, where it feeds on invertebrates. Male's call, a guttural *wrocc*, is made from within a chamber of mud, usually beneath rocks and leaf litter. About 100 eggs are laid in a foam nest, and tadpoles start to develop into metamorphlings after 7 weeks.

Scott Eipper

Pugh's Mountain Frog ■ *Philoria pughi* TL 30mm

DESCRIPTION Brown or yellow above. Legs and hips usually reddish-brown, with prominent dark stripe from nostril, through eye and on to shoulder. Typically has large dark spot on lower flank, in front of hindlimbs. Underside yellow with darker markings. **DISTRIBUTION** Found in a few upland locations in northern NSW near Washpool

National Park. **HABITS AND HABITAT** Cathemeral. Lives in seepages in wet sclerophyll forests and rainforests, where it feeds on invertebrates. Male's call, a guttural *wrocc*, sometimes with a low, descending growl, is made from inside a chamber of mud, usually beneath rocks and leaf litter. Reproductive habits little known, but thought to be similar to those of the Sphagnum Frog (see opposite).

Angus McNab

Richmond Range Sphagnum Frog ■ *Philoria richmondensis* TL 28mm

DESCRIPTION Reddish-brown to dark brown or yellow above, with distinct pale-edged, dark stripe from snout to above shoulder. Body smooth. Bright yellow to grey below with indistinct markings.

DISTRIBUTION Found in a few upland locations on Richmond Range, NSW.

HABITS AND HABITAT Cathemeral. Lives in seepages and gutters in rainforests, where it feeds on invertebrates. Male's call, a guttural *wrocc*, is made from within a chamber of mud, usually beneath rocks and leaf litter. Eggs laid in a foam nest, and once hatched tadpoles start to develop into metamorphlings after 10 weeks.

Angus McNab

Sphagnum Frog ■ *Philoria sphagnicolus* TL 37mm

DESCRIPTION Almost any shade of grey or brown above, but can be reddish, orange or yellow, usually with dark stripe from rear of eye to above shoulder that continues down on to lower flank. Often has darker mottling, and occasionally 'V'- or 'W'-shaped black marks, centring on midline. Body smooth. White to pale brown below, with or without heavy brown mottling. **DISTRIBUTION** Found in a few upland locations of north-east NSW, from Glen Innes to Comboyne. **HABITS AND HABITAT** Cathemeral. Lives in beds of

sphagnum moss, seepages in wet sclerophyll forests and rainforests, where it feeds on invertebrates. Male's call, a guttural *wrocc* repeated in short bursts, is made from within a chamber of mud, usually beneath rocks and leaf litter. About 50 eggs are laid in a foam nest, and following hatching tadpoles start to develop into metamorphlings after 8–11 weeks, depending on the temperature.

Grant Webster

Ornate Burrowing Frog ▪ *Platyplectrum ornatum* TL 49mm

Scott Eipper

DESCRIPTION Upperparts very variable in colouration and patterning – a combination of mottling, stripes and flecks, usually with dark stripe from rear of eye to above shoulder. Body smooth and rounded. Cream to white below, usually without markings. **DISTRIBUTION** Found across northern and eastern Australia, from Port Hedland, WA, to western Sydney, NSW. **HABITS AND HABITAT** Nocturnal. Inhabits dry woodland,

Scott Eipper

brigalow, black-soil plains and savannah woodland, where it lives in burrows below the ground and emerges at night, usually after rain, to feed on invertebrates. Male's call, a short *unk* repeated every couple of seconds, is made while floating in still water. About 1,000 eggs are laid in still pools. Length of time as tadpoles largely depends on the temperature, but is usually 3–6 weeks.

Spencer's Burrowing Frog ▪ *Platyplectrum spenceri* TL 52mm

DESCRIPTION Very variable in colouration and patterning above. Typically a combination of mottling, stripes and flecks, usually with dark stripe from rear of eye to above shoulder. Body smooth and rounded. Cream to white below, usually without markings. **DISTRIBUTION** Found across western and central Australia, from Port Hedland, WA, to Thargomindah, Qld. **HABITS AND HABITAT** Nocturnal. Inhabits

Scott Eipper

dry woodland, deserts, black-soil plains and savannah, where it lives in burrows below the ground, emerging at night, usually after rain, to feed on invertebrates. Male's call, a short *errk* repeated every second or so, is made while floating in still water. About 900 eggs are laid in still pools. Length of time as tadpoles largely depends on the temperature, but is usually 3–6 weeks.

> **MICROHYLIDAE (NARROW-MOUTHED FROGS)**
> Australian members of this moderately large family of small frogs are typically found
> in relatively wet forests in the north of the country. Of the many subfamilies in the
> Microhylidae, only the Asterophryinae occurs in Australia. Adult frogs lay their eggs
> on land, usually beneath rocks and logs, from which fully formed frogs hatch.

Northern Territory Frog ▪ *Austrochaperina adelphe* TL 22mm

DESCRIPTION Reddish to yellowish-brown above, often with darker longitudinal
markings and numerous white flecks covering dorsum. Usually a dark stripe from snout
extending to past shoulder. Some individuals have grey flanks. White to cream below.
DISTRIBUTION Found in Top End
region of NT, including offshore
islands. **HABITS AND HABITAT**
Nocturnal. Occurs in swamps,
monsoon forests and wet areas
of open forests, where it feeds on
invertebrates. Male's call, a short,
repeated beep similar to a warning
from a smoke detector, is made from
beneath leaf litter. Lays about 12
eggs, which are guarded by male
until they hatch into small frogs.

Ryan Francis

Fry's Frog ▪ *Austrochaperina fryi* TL 40mm

DESCRIPTION Reddish, to yellowish-brown, to chocolate above, usually with dark stripe
from snout, extending along above tympanum and terminating near shoulder. Often
has darker flecking, but can be immaculate, and occasionally has a thin vertebral stripe.
Underside yellow; some individuals
have orange around armpit and
groin. **DISTRIBUTION** Found in
north-east Qld, from Lake Barrine to
Rossville. **HABITS AND HABITAT**
Nocturnal. Occurs in rainforests,
where it feeds on invertebrates.
Male's call, a short, beeping whistle
comprising about 8 rapidly repeated
notes in a burst, is made from
beneath leaf litter. Lays about 9 eggs,
which are guarded by male until
they hatch into small frogs.

Scott Eipper

Slender Whistling Frog ■ *Austrochaperina gracilipes* TL 22mm

DESCRIPTION Reddish, to yellowish-brown, to grey above, often with darker and lighter flecking, and usually with dark stripe from snout, extending along side to past shoulder. Some individuals have a dark streak beginning midway along back over hips. Grey below with white and black flecking. Most individuals have orange around groin that extends up on to lower flanks. **DISTRIBUTION** Found in north-east Qld on Cape York from Cooktown to tip. **HABITS AND HABITAT** Nocturnal. Occurs in vine forests and rainforest pockets along creek lines, where it feeds on invertebrates. Male's call, a beeping whistle that pulses about every second, is made from beneath leaf litter. Lays about 8 eggs, which are guarded by male until they hatch into small frogs.

Anders Zimny

White-browed Whistling Frog ■ *Austrochaperina pluvialis* TL 30mm

DESCRIPTION Reddish to yellowish-brown, chocolate or blackish above, occasionally with white flecking. Snout and area across tops of eyes often has a white to yellow stripe that extends on to body across shoulder. Below this a dark stripe extends from snout along flanks to groin. Yellow to cream below. **DISTRIBUTION** Found in north-east Qld, from Ingham to Mossman. **HABITS AND HABITAT** Nocturnal. Inhabits rainforests, where it feeds on invertebrates. Male's call, a series of rapid, whistling pulses that lasts less than a second, is made from beneath leaf litter. Lays about 12 eggs, which are guarded by male until they hatch into small frogs after about 28 days.

Angus McNab

Robust Whistling Frog ■ *Austrochaperina robusta* TL 33mm

DESCRIPTION Reddish to yellowish-brown or chocolate above, often with darker flecking, but can be immaculate. Some individuals have darker flanks extending along body, and there may be a thin vertebral stripe. Occasionally has a light-coloured brow, similar to that of the White-browed Whistling Frog (see opposite). Yellow to cream below. **DISTRIBUTION** Found in north-east Qld, from Mt Elliot to Cairns. **HABITS AND HABITAT** Nocturnal. Inhabits rainforests, where it feeds on invertebrates. Male's call, about 8 whistling pulses that last around 2 seconds, is made from beneath leaf litter. Lays about 11 eggs, which are guarded by male until they hatch into small frogs.

Scott Eipper

Tapping Nursery Frog ■ *Cophixalus aenigma* TL 24mm

DESCRIPTION Typically orange, grey-brown or blackish above, and can be reddish to yellowish-brown or chocolate, often with darker flecking, but may be immaculate. With or without tubercles. Some individuals have thin vertebral stripe. Cream to grey below, with or without markings, and normally with yellow or orange markings beneath feet and limbs, extending up on to lower flanks. Upper iris often pale blue or green. **DISTRIBUTION** Found in north-east Qld, from Mt Lewis to Thornton Peak, above 700m altitude. **HABITS AND HABITAT** Nocturnal. Occurs in rainforests and boulder fields, where it feeds on invertebrates. Male's call, a series of tapping clicks that last about 3–6 seconds, is made from leaf litter or while in low vegetation. Lays about 12 white eggs, which are guarded by male until they hatch into small frogs.

Scott Eipper

Southern Ornate Nursery Frog ■ *Cophixalus australis* TL 28mm

DESCRIPTION Colouration very variable. Most common variants are cream, beige or greyish, with or without a dark, 'W'-shaped marking between shoulders. Usually has other dark streaks over shoulders and hips. Other individuals are yellowish to dark brown with darker flecking. Randomly scattered tubercles. Cream to grey below, with or without darker markings. **DISTRIBUTION** Occurs in north-east Qld, from Malbon Thompson

Range to Paluma. **HABITS AND HABITAT** Nocturnal. Found in rainforests and gardens, where it feeds on invertebrates. Male's call is a drawn-out bleat, which is made from a few centimetres up to 3m above the ground, while sitting on a leaf, branch or perched between two vertical branches. Lays about 14 white eggs, which are guarded by male until they hatch into small frogs.

Buzzing Nursery Frog ■ *Cophixalus bombiens* TL 17mm

DESCRIPTION One of Australia's smallest frogs. Colouration extremely variable, the most common variants being cream, beige or grey, usually with dark streaks over shoulders and hips. Other individuals yellowish to dark brown, often with darker flecking. Randomly

scattered tubercles. Cream to grey below, with or without darker markings. **DISTRIBUTION** Occurs in north-east Qld, restricted to Mossman Gorge to Windsor tableland area. **HABITS AND HABITAT** Nocturnal. Found in rainforests, where it feeds on invertebrates. Male's call, an insect-like buzz, is made from beneath leaf litter. Lays about 6 white eggs, which are guarded by male until they hatch into small frogs.

Beautiful Nursery Frog ■ *Cophixalus concinnus* TL 26mm

DESCRIPTION Orange, grey or purplish above, occasionally blotched with dark spots. With or without tubercles. Cream to grey below, with large red to orange markings. Most individuals have yellow or orange markings beneath feet and limbs extending up on to lower flanks. Upper iris often pale blue or green. **DISTRIBUTION** Found in north-east

Qld on top of Thornton Peak above 1,100m altitude. **HABITS AND HABITAT** Nocturnal. Inhabits rainforests and vine scrub, where it feeds on invertebrates. Male's call, a *creeeeeeeee* that last about 2 seconds, is made from low shrubs 50cm–2m above the ground. Lays 17 white eggs, which are thought to be guarded by male until they hatch into small frogs.

McIlwraith Nursery Frog ■ *Cophixalus crepitans* TL 14mm

DESCRIPTION One of Australia's smallest frogs. Usually rusty-orange with dark brown to black markings, and normally has dark streaks over shoulders and hips. Other individuals yellowish to dark brown, often with darker flecking. Randomly scattered tubercles across upper body. Cream to grey below, with or without darker markings. Very similar to the Cape York Nursery Frog (see p. 55), with which it may be conspecific, and only differs by

limb length and call pitch. **DISTRIBUTION** Occurs in north-east Qld, from near Peach Creek in McIlwraith Range. **HABITS AND HABITAT** Nocturnal. Found in rainforests, where it feeds on invertebrates. Male's call, a series of rapidly repeated clicks, is made from above the ground, while perched on vegetation. Reproduction unknown, but considered to be similar to other species in the genus.

49

Northern Tapping Nursery Frog ▪ *Cophixalus exiguus* TL 16mm

DESCRIPTION Beige or greyish above, usually with 'V'-shaped markings running down back from between shoulders. Other dark streaks often present, including a postocular stripe. Some individuals are plain brown, often with darker flecking. Body smooth, with or without low tubercles. Cream to grey below, with or without darker markings.

DISTRIBUTION Occurs in north-east Qld, from Big Tableland to Mt Findlay in northern wet tropics. **HABITS AND HABITAT** Nocturnal. Found in rainforests and gardens, where it feeds on invertebrates. Male's call is similar to the sound of marbles being shaken in a bottle, and is made from on the ground or in vegetation. Lays about 9 white eggs, which are thought to be guarded by male until they hatch into small frogs.

Hinchinbrook Island Nursery Frog
▪ *Cophixalus hinchinbrookensis* TL 26mm

DESCRIPTION Colouration extremely variable. Most common variants are cream, beige or greyish, with or without a dark 'W'-shaped marking between shoulders. Other dark streaks are usually present over shoulders and hips. Some individuals are yellowish to dark brown, often with darker flecking. Randomly scattered tubercles are present. Cream to grey

below, with or without white markings. **DISTRIBUTION** Occurs in north-east Qld on Hinchinbrook Island, 300m asl. **HABITS AND HABITAT** Nocturnal. Inhabits rainforests, heaths and rocky creek lines, where it feeds on invertebrates. The male's call, a short *beeep*, is given from above ground, while sitting on a leaf or branch. Lays about 8 white eggs, which are guarded by male until they hatch into small frogs.

Hosmer's Nursery Frog ■ *Cophixalus hosmeri* TL 15mm

DESCRIPTION Colouration very variable. Can be pale brown, beige, grey, reddish to black above, with or without pattern. Body can be smooth, heavily ridged or covered in tubercles. Underside colouration is as variable as dorsum. Short snout gives it a stout appearance. **DISTRIBUTION** Occurs in north-east Qld, on Mt Lewis and Mt Spurgeon, at 800m asl. **HABITS AND HABITAT** Nocturnal. Found in rainforests, where it feeds on invertebrates. Male's call is similar to the sound of marbles being shaken in a bottle, but can vary, and is given from above the ground, while perched in low vegetation and in palm axils. Lays about 6 white eggs, which are guarded by male until they hatch into small frogs.

Scott Eipper

Creaking Nursery Frog ■ *Cophixalus infacetus* TL 18mm

DESCRIPTION Typically orange, grey-brown or blackish above, but can be reddish to yellowish-brown or chocolate, often with darker flecking. Some individuals have a broad vertebral stripe. Skin can be smooth, or with raised tubercles forming small ridges. Cream to grey below with white stippling and spotting. **DISTRIBUTION** Occurs in north-east Qld, from Gordonvale to Cardwell. **HABITS AND HABITAT** Nocturnal. Found in rainforests, along creek lines and other waterways, where it feeds on invertebrates. Male's call sounds like a creaking door opening, and is given from within leaf litter or while in low vegetation. Lays about 10 white eggs, which are guarded by male until they hatch into small frogs.

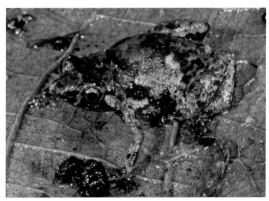

Scott Eipper

Kutini Boulder Frog ■ *Cophixalus kulakula* TL 47mm

DESCRIPTION Brown or grey to purple-grey above, usually with dark canthal streak that extends through eye to shoulder; some individuals have black flecking and blotches. Skin can be smooth or have raised tubercles. Upper iris gold flecked. White to grey below with white stippling and spotting, and groin and inner thigh reddish-orange. **DISTRIBUTION**

Occurs in north-east Qld, at Mt Tozer and Tor Hill near Iron Range. **HABITS AND HABITAT** Nocturnal. Found in granite boulder fields in gullies in rainforests, where it feeds on invertebrates. Male's call, a short *wrarrk*, is made from the leaf litter or while in low vegetation. The single reproductive record was a clutch of 47 eggs.

Male

Female

Mount Elliott Nursery Frog ■ *Cophixalus mcdonaldi* TL 20mm

DESCRIPTION Yellowish-beige or greyish above, with or without dark 'W'- or 'V'-shaped markings on back. Usually has other dark streaks on flanks and some mottling. Randomly scattered tubercles. Cream to grey below, with or without white markings. **DISTRIBUTION** Found in north-east Qld, on Mt Elliot, above 750m elevation. **HABITS AND HABITAT** Nocturnal. Inhabits rainforests and rocky creek lines near summit, where it feeds on invertebrates. Male's call is a drawn-out *creeeeeek* made from above the ground while sitting perched on a leaf or branch. Lays about 17 white eggs, which are guarded by male until they hatch into small frogs.

Anders Zimny

Mountain-top Nursery Frog ■ *Cophixalus monticola* TL 22mm

DESCRIPTION Colouration extremely variable. Can be pale brown, beige, grey, reddish to black above, with or without pattern. Body can be smooth, heavily ridged or covered in tubercles. Underside colouration as variable as dorsum. Short snout gives it a stout appearance. **DISTRIBUTION** Found in north-east Qld, on Mt Lewis, above 1,000m asl. **HABITS AND HABITAT** Nocturnal. Occurs in rainforests, where it feeds on invertebrates. Male's call is a drawn-out *creeeeeek* made from above the ground while perched in low vegetation and in palm axils. Lays about 10 white eggs, which are guarded by male until they hatch into small frogs.

Scott Eipper

Neglected Nursery Frog ■ *Cophixalus neglectus* TL 29mm

DESCRIPTION Colouration variable, from brown to reddish-orange, without pattern. Body smooth, although it is often infected with a mite that produces small, raised lumps. Yellow, orange or cream below. **DISTRIBUTION** Found in north-east Qld, on Mt Bellenden Ker and Mt Bartle Frere, above 1,150m asl. Range may have contracted due to climate change as there are historical records from lower elevations. **HABITS AND**

Anders Zimny

HABITAT Nocturnal. Inhabits rainforests, where it feeds on invertebrates. Two populations, which are different genetically and have different calls. Bellenden Ker population's call is a *wreeek*, while Bartle Frere population's call is a *squeeeelch*. Males of both populations call from leaf litter or while in low vegetation. Lays about 14 white eggs, which are guarded by male until they hatch into small frogs.

Northern Ornate Nursery Frog ■ *Cophixalus ornatus* TL 29mm

DESCRIPTION Colouration very variable. Common variants are cream, beige or greyish, with or without dark, 'W'-shaped marking between shoulders. Other dark streaks usually present over shoulders and hips. Individuals can also be yellowish to dark brown, typically with darker flecking. Randomly scattered tubercles. Cream to grey below, with

Scott Eipper

or without darker markings. **DISTRIBUTION** Found in north-east Qld, from Mt Lewis to Mt Bartle Frere. **HABITS AND HABITAT** Nocturnal. Inhabits rainforests and gardens, where it feeds on invertebrates. Male's call is a drawn-out *weeeeeeercc*, given while perched on a leaf or branch a few centimetres up to 2m above the ground. Lays about 16 white eggs, which are guarded by male until they hatch into small frogs up to 28 days after being laid.

Golden-capped Nursery Frog ■ *Cophixalus pakayakulangun* TL 50mm

DESCRIPTION The largest Australian microhylid. Brown, grey or purple-grey above, usually with dark canthal streak that extends through eye to tympanum, and with dull golden triangle running across tops of eyes and to nostrils. Some individuals have black flecking. Skin can be smooth or have raised tubercles. White to grey below with white stippling and spotting, and groin is plain. **DISTRIBUTION** Found in north-east Qld, at Stanley Hill on Cape York Peninsula. **HABITS AND HABITAT** Nocturnal. Occurs in granite boulder fields in gullies within rainforests, where it feeds on invertebrates. Male's call is like a rapidly bouncing marble, and is given from leaf litter or while in low vegetation. Lays a clutch of about 46 eggs.

Anders Zimny

Cape York Nursery Frog ■ *Cophixalus peninsularis* TL 19mm

DESCRIPTION Typically rusty-orange above with dark brown to black markings and dark streaks over shoulders and hips. Some individuals yellowish to dark brown, often with darker flecking. Randomly scattered tubercles across upper body. Cream to grey below, with or without darker markings. Very similar to the McIlwraith Nursery Frog (see p. 49), with which it may be conspecific, only differing by limb size and call pitch. **DISTRIBUTION** Occurs in north-east Qld, from Rocky River Scrub at 540m elevation in McIlwraith Range. **HABITS AND HABITAT** Nocturnal. Found in rainforests, where it feeds on invertebrates. Males call from above the ground, while perched on vegetation. Reproduction unknown, but thought to be similar to that of other species in the genus.

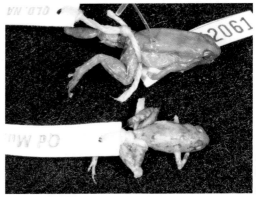

C. peninsularis *upper*, C. crepitans *below*

Conrad Hoskin

Blotched Boulder Frog ■ *Cophixalus petrophilus* TL 30mm

DESCRIPTION Yellow to pale cream above, blotched with dark spots, and usually with a dark canthal streak that extends through eye to ear. Skin has raised tubercles. Upper

iris gold flecked. White to grey below, and groin and inner thigh reddish-orange. **DISTRIBUTION** Found in north-east Qld, at Cape Melville. **HABITS AND HABITAT** Nocturnal. Recently described from granite boulder fields surrounded by rainforest, where it feeds on invertebrates. Male's call consists of a series of rapidly repeated clicks, given from within rock crevices. Reproductive habits unknown.

Black Mountain Boulder Frog ■ *Cophixalus saxatilis* TL 47mm

DESCRIPTION Male yellow-brown or grey above, mottled with black. Female bright yellow, often with dark smudging on back. Flanks bright yellow. Usually a dark temporal streak that extends from eye to shoulder. Body smooth or with low tubercles. Upper iris

gold flecked. Cream below with grey stippling and spotting, and groin and inner thigh are orange. **DISTRIBUTION** Found in north-east Qld, at Black Mountain, south of Cooktown. **HABITS AND HABITAT** Nocturnal. Inhabits granite boulder fields in gullies, where it feeds on invertebrates. Male's call, a short *wrarrk*, is made from within rock crevices. Produces clutch of about 25 eggs.

Cape Melville Boulder Frog ■ *Cophixalus zweifeli* TL 40mm

DESCRIPTION Various tones of brown to rusty-orange above, with dark streak that extends from nostril, becoming broken over shoulder and forming blotches on lower flanks. Many individuals have a diffuse golden mark forming triangle between eyes and nostrils. Body smooth or with low tubercles. Upper iris gold flecked. Cream below, with grey stippling and spotting, and groin and inner thigh are orange-red. **DISTRIBUTION** Occurs in north-east Qld, at Cape Melville. **HABITS AND HABITAT** Nocturnal. Inhabits crevices in granite boulder fields, and feeds on invertebrates. Male's call is a short, cat-like *meow*, given from within a rock crevice. Reproductive habits unknown.

H. B. Hines/DES

> **MYOBATRATCHIDAE (GROUND FROGS)**
> Unique to Australia and New Guinea, this group of small to medium-sized frogs is
> often referred to as ground frogs. Species within the family vary greatly in habits
> and habitat, and exhibit some of the world's most diverse adaptations to get their
> young from egg to neonate frog. Most have aquatic free-swimming larvae (tadpoles),
> while some have young that fully develop within the eggs. The inclusion of the
> *Rheobatrachus* and *Mixophyes* genera in the Myobatrachidae is likely to change.

White-bellied Frog ▪ *Anstisia alba* TL 25mm

DESCRIPTION Grey to brownish-pink above, with small bumps and ridges on back,
usually dark grey to black. Skin surface granular. White below. Recently split from the

genus *Geocrinia*. **DISTRIBUTION** Occurs
in southwestern WA, from Karridale to
Witchcliffe. **HABITS AND HABITAT**
Nocturnal. Lives in open woodland and
swamps, typically along seepages, and feeds on
invertebrates. Call is a series of rapid clicks.
About 15 eggs are laid in a burrow, which
hatch and remain in the gelatinous mass in
which they are laid. Tadpoles metabolize the
yolk they hatch with and do not feed, turning
into frogs after about 8 weeks.

Walpole Frog ▪ *Anstisia lutea* TL 25mm

DESCRIPTION Olive to dark brown above, with dark mottling. Back has small bumps and
ridges and skin surface is granular. Yellowish to cream below, with dark speckling; throat
of calling males blackish. Recently split from the genus *Geocrinia*. **DISTRIBUTION** Found
in southwestern WA, from Walpole to Nornalup. **HABITS AND HABITAT** Nocturnal.

Inhabits wet, closed woodland,
typically along seepages, and feeds
on invertebrates. Call is a series
of single clicks, repeated in bursts.
About 25 eggs are laid in depressions
alongside streams and creeks,
beneath vegetation mats. After
hatching, the tadpoles remain in the
gelatinous mass in which they are
laid. They metabolize the yolk they
hatch with and do not feed, turning
into frogs after about 8 weeks.

Karri Frog ■ *Anstisia rosea* TL 25mm

DESCRIPTION Blackish to dark brown above, with dark mottling. Back has small bumps and ridges, and skin surface is granular. Rose-pink below; throat of calling males blackish. Recently split from the genus *Geocrinia*. **DISTRIBUTION** Found in southwestern WA, from Margaret River to Walpole. **HABITS AND HABITAT** Nocturnal. Lives in Jarrah and Karri woodland and swamps, typically along seepages, and feeds on invertebrates. Call is a single click repeated every half second. About 25 eggs are laid in depressions alongside streams and creeks beneath vegetation mats, and after hatching the tadpoles remain in the gelatinous mass in which they are laid. They metabolize the yolk they hatch with and do not feed, turning into frogs after about 9 weeks.

Grant Webster

Yellow-bellied Frog ■ *Anstisia vitellina* TL 25mm

DESCRIPTION Greyish-white to brownish-orange above. Back has small bumps and ridges that are usually dark grey to black. Skin surface granular. Bright yellow below, with some white speckling towards rear. Recently split from the genus *Geocrinia*.

DISTRIBUTION Found in south-west WA, on the slopes of Spearwood Creek. **HABITS AND HABITAT** Nocturnal. Lives in seepages in dense woodland. Feeds on invertebrates. Call is a series of rapid clicks. About 15 eggs are laid in a depression beneath vegetation mats. They hatch and remain in the gelatinous mass in which they are laid. Tadpoles metabolize the yolk they hatch with and do not feed, turning into frogs after about 8 weeks.

Grant Webster

Northern Sandhill Frog ■ *Arenophryne rotunda* TL 39mm

DESCRIPTION Body stocky, with short legs and toes, and a rounded face. Colouration off-white to cream, but can be greyish or brownish, and variably spotted above with red, green and brown. **DISTRIBUTION** Found in western WA, from Shark Bay and Edel Land; also on Dirk Hartog Island. **HABITS AND HABITAT** Nocturnal. Occurs in sand dunes, where

it spends the majority of its time in a burrow a short distance below the surface, emerging on damp or rainy nights to feed on insects, particularly ants. Call is a short squelch made while underground in the burrow. Does not require water for reproduction, laying its 4–11 eggs deep in moist sand, with young frogs hatching directly from eggs after 9 weeks.

Southern Sandhill Frog ■ *Arenophryne xiphorhyncha* TL 36mm

DESCRIPTION Body stocky, with short legs and toes, and a pointed face. Colouration off-white to cream, but can be greyish or brownish, and variably spotted above with red, green and brown. **DISTRIBUTION** Found in western WA, from Kalbarri to Cooloomia Station. **HABITS AND HABITAT** Nocturnal. Occurs in sand dunes, where it spends the majority

of its time in a burrow a short distance below the surface, emerging on damp or rainy nights to feed on insects, particularly ants. Call is a short squelch made while underground in the burrow. Does not require water for reproduction, laying its 4–11 eggs deep in moist sand, with young frogs hatching directly from eggs after about 9 weeks.

Hip Pocket Frog ■ *Assa darlingtoni* TL 20mm

DESCRIPTION Grey to brown or blackish above, sometimes mottled or with flecks, and some individuals have 'V'-shaped mark over shoulders and hips. White below with brown markings. Not distinguished morphologically from Mount Wollumbin Hip Pocket Frog (see below). **DISTRIBUTION** Occurs in eastern Australia, from the Sunshine Coast, Qld, to Coffs Harbour, NSW. Does not occur on Mt Warning, NSW. **HABITS AND HABITAT** Nocturnal. Found in rainforests and wet forests, where it feeds on invertebrates. Male's call

is a series of 4–14 fast-paced *wrucs* produced from beneath leaf litter on the forest floor. Call has on average more notes than Mount Wollumbin Hip Pocket Frog. Lays about 15 eggs. After hatching, tadpoles are collected by the male, which uses two hip pouches to rear them through metamorphosis. Fully formed young emerge from pouch after tadpole stage of 60–81 days.

Scott Eipper

Mount Wollumbin Hip Pocket Frog ■ *Assa wollumbin* TL 20mm

DESCRIPTION Grey to brown or blackish above, sometimes mottled or with flecks; some individuals have a 'V'-shaped mark over shoulders and hips. White below with brown markings. Not distinguishable morphologically from the Hip Pocket Frog (see above). **DISTRIBUTION** Found only on Mt Warning, NSW. **HABITS AND HABITAT**

Nocturnal. Found in rainforests and wet forests, where it feeds on invertebrates. Male's call a series of 4–14 fast-paced *wrucs* produced from beneath leaf litter on the forest floor. Call has on average fewer notes than that of the Hip Pocket Frog. Lays about 15 eggs. After hatching, tadpoles are collected by male, which uses two hip pouches to rear them through metamorphosis. Fully formed young emerge from pouch after tadpole stage of 60–81 days.

Ian Bool

Bilingual Froglet ▪ *Crinia bilingua* TL 25mm

DESCRIPTION Pattern, colouration and texture of skin vary widely from brown, to light grey, to almost black, with or without flecks and stripes. Back can be smooth, without bumps or protrusions, or extremely warty, and every variation in between. Underside white

to black, with or without markings. DISTRIBUTION Found from Gulf region of NT, across to Kimberley, WA. HABITS AND HABITAT Nocturnal. Inhabits savannah, swamps and agricultural areas, where it feeds on invertebrates. Male calls while floating in a waterbody, emitting a short chirp followed by a drawn-up rattle. About 180 eggs are laid in still pools, and tadpoles start to develop into metamorphlings within 4 weeks.

Desert Froglet ▪ *Crinia deserticola* TL 18mm

DESCRIPTION Usually olive to grey or pale brown, with or without darker mottling. These markings can form an inverted 'V' aligned down the spine. Back can be smooth, without bumps or protrusions, or extremely warty, and every variation in between. Underside pale and granular. DISTRIBUTION Found from Lake Argyle, WA, through northern and inland

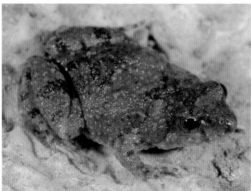

Australia, to far north-west NSW, and across to near Childers, Qld. HABITS AND HABITAT Nocturnal. Inhabits savannah, brigalow, black-soil plains, grassland, swamps and agricultural areas, where it feeds on invertebrates. Male's call, a short *beep*, is given while sitting around a waterbody. About 150 eggs are laid in still pools, and tadpoles start to develop into metamorphlings within 4 weeks.

Kimberley Froglet ■ *Crinia fimbriata* TL 19mm

DESCRIPTION Typically reddish to grey or pale brown above, with or without darker mottling, and with conspicuous white tubercles. Pale brown to black below with bright white flecking. Legs are banded and digits have distinct fringes in males. **DISTRIBUTION** Found in WA, at Mitchell Plateau and Prince Regent River. **HABITS AND HABITAT** Nocturnal. Inhabits rocky gorges, where it feeds on invertebrates. The species was recently described and very little is known about its reproductive biology, other than that it breeds in rock pools.

Northern Flinders Ranges Froglet ■ *Crinia flindersensis* TL 26mm

DESCRIPTION Pattern, colouration and texture of skin vary widely from brown, to light grey, to dark brown, with or without flecks and stripes. Back can be smooth, without bumps or protrusions, or extremely warty, and every variation in between. Underside white with black or dark grey markings. One of two *Crinia* species with a hidden tympanum. **DISTRIBUTION** Found in SA, in Flinders Ranges between Wilkatana Station and Mt Freeling Station. **HABITS AND HABITAT** Nocturnal. Lives in gorges with streams and creeks, where it is most common in fast-flowing sections, and feeds on invertebrates. Recently found to be distinct from the Southern Flinders Range Froglet (see p. 67) on the basis of tadpole morphology, distribution and genetics. Male's call sounds like *craaaaar k*, and is made while floating in water or next to it. About 100 eggs are laid in pools.

Tschudi's Froglet ▪ *Crinia georgiana* TL 45mm
(Quacking Froglet)

DESCRIPTION Pattern, colouration and texture of skin above can vary widely from yellow, to light grey, to almost black, with or without flecks and stripes. Back can be smooth, without bumps or protrusions, or extremely warty, and every variation in between. White to orange-pink below, with black throat in males. Usually has red flush

on tops of eyes and red over groin and limbs. Forelimbs in males are very robust. **DISTRIBUTION** Found in south-west WA, from Gingin to Cape Le Grande. **HABITS AND HABITAT** Inhabits woodland and granite outcrops around slow-moving or still watercourses, where it feeds on invertebrates. Call sounds like a quacking duck's. About 115 eggs are laid singly, sometimes attached to plant material, and tadpoles start to develop into metamorphlings after 4 weeks.

Clicking Froglet ▪ *Crinia glauerti* TL 25mm

DESCRIPTION Pattern, colouration and texture of skin above vary widely from brown to light grey or almost black, with or without flecks and stripes. Back can be smooth, without bumps or protrusions, or extremely warty, and every variation in between. White to black

below, with markings in females, while calling males are black to dark grey with fine white crucifix marking. **DISTRIBUTION** Occurs in south-west WA, from Moore River to Pallinup River. **HABITS AND HABITAT** Inhabits woodland and forests around slow-moving or still watercourses, where it feeds on invertebrates. Call sounds like marble bouncing on a hard surface. About 100 eggs are laid singly and sit on the bottom of the pool. Tadpoles start to develop into metamorphlings after 12 weeks.

Squelching Froglet ▪ *Crinia insignifera* TL 25mm

DESCRIPTION Generally uniform brown, grey or black above, but can be a mix of these colours, with darker bars on legs, and normally with darker triangular patch between eyes. Underparts paler, blotched and streaked with darker markings, and with raised nodules on belly. **DISTRIBUTION** Found in Rottnest Island and Swan Coastal Plain, southwestern WA, from around Jurien to Busselton. **HABITS AND HABITAT** Inhabits moist areas

such as rivers, streams and swamps, where it feeds on invertebrates. Male's call sounds like *weee weee weee*, repeated frequently, or like a finger being dragged across a balloon, and is made from emergent vegetation, flooded grassland and pond edges. About 140 eggs are laid in pools of still water, and tadpoles take about 15 weeks to metamorphose into frogs.

Scott Eipper

Moss Froglet ▪ *Crinia nimba* TL 30mm

DESCRIPTION Colouration and texture of skin vary widely from tan to light grey or dark brown, with dark stripes and blotches, and typically with dark 'V'-shaped marking centred behind eyes. Back can be smooth, without bumps or protrusions, or extremely warty, and every variation in between. Brown below with white flecking. **DISTRIBUTION**

Occurs in a few parts of southern Tas, particularly in upland areas. **HABITS AND HABITAT** Diurnal. Inhabits moorland, cloud forests and rainforests, where it feeds on invertebrates. Also known as *Bryobatrachus nimbus* in some texts. The call, a short click, is repeated frequently. About 12 eggs are laid in a gelatinous mass inside a nest chamber. After hatching tadpoles do not feed, taking around 11–14 months from oviposition to completing metamorphosis.

Angus McNab

Eastern Sign-bearing Froglet ▪ *Crinia parinsignifera* TL 23mm

DESCRIPTION Pattern, colouration and texture of skin vary widely from yellow, to light grey, to almost black, with or without flecks and stripes. Back can be smooth, without bumps or protrusions, or extremely warty, and every variation in between. Underside white with black flecking. **DISTRIBUTION** Found from Rockhampton, Qld, to south-east SA. **HABITS AND HABITAT** Lives around slow-moving or still watercourses, including

artificial dams, lakes and ornamental pools, where it feeds on invertebrates. Male's call is a rapid clicking similar to the sound of a ratchet, and is repeated frequently from emergent vegetation, flooded grassland and pond edges. About 130 eggs are laid singly, sometimes attached to plant material, and after hatching tadpoles start to develop into metamorphlings after 23 weeks.

Bleating Froglet ▪ *Crinia pseudinsignifera* TL 30mm

DESCRIPTION Pattern, colouration and texture of skin vary greatly from brown, to light grey, to almost black, with or without flecks and stripes. Back can be smooth, without bumps or protrusions, or extremely warty, and every variation in between. White to black

below, with or without markings. **DISTRIBUTION** Found in south-west WA, from Kalbarri to Israelite Bay. **HABITS AND HABITAT** Inhabits woodland with granite outcrops around slow-moving or still watercourses, where it feeds on invertebrates. Call sounds like bleating sheep. About 120 eggs are laid singly, which sit on the bottom of a pool, and tadpoles start to develop into metamorphlings 12 weeks after hatching.

Northern Froglet ▪ *Crinia remota* TL 18mm

DESCRIPTION Pattern, colouration and texture of skin above can vary widely, from brown, to light grey, to almost black, with or without flecks and stripes. Back can be smooth, without bumps or protrusions, or extremely warty, and every variation in between. White to black below, with or without markings. **DISTRIBUTION** Found from Townsville, Qld, to NT/ WA border. **HABITS AND HABITAT** Lives in savannah, swamps and agricultural areas, where it feeds on invertebrates. Male's call, a rapidly repeated clicking, is given while floating in a waterbody. Eggs are laid in still pools, and tadpoles start to develop into metamorphlings within 6 weeks of hatching.

Ryan Francis

Southern Flinders Ranges Froglet ▪ *Crinia riparia* TL 25mm

DESCRIPTION Pattern, colouration and texture of skin vary widely from brown, to light grey, to dark brown, with or without flecks and stripes. Back can be smooth, without bumps or protrusions, or extremely warty, and every variation in between. Underside white with black or dark grey markings. One of two *Crinia* species with a hidden tympanum. **DISTRIBUTION** Found in SA, in Flinders Ranges between Warren Gorge and Mambray Creek. **HABITS AND HABITAT** Lives in gorges with streams and creeks, usually in fast-flowing sections. Feeds on invertebrates. Male's call sounds like *ick ick ick chrrrk*, and is given while floating in or next to a waterbody. About 100 eggs are laid in pools.

Adam Elliott

Common Froglet ■ *Crinia signifera* TL 25mm

DESCRIPTION Pattern, colouration and texture of skin above can vary widely from yellow, to light grey, to almost black, with or without flecks and stripes. Back can be smooth, without bumps or protrusions, or extremely warty, and every variation in between. Underparts white with black flecking. **DISTRIBUTION** Occurs from south-east Qld to south-east SA, including Tas. **HABITS AND HABITAT** Cathemeral. Very common across its range. Found around slow-moving or still watercourses, including artificial dams,

lakes and ornamental pools, where it feeds on invertebrates. Male's call is a rapid clicking similar to the sound of a ratchet, and is repeated frequently from emergent vegetation, flooded grassland and pond edges. About 400 eggs are laid singly, sometimes attached to plant material, and tadpoles start to develop into metamorphlings within 6 weeks of hatching.

Sloane's Froglet ■ *Crinia sloanei* TL 22mm

DESCRIPTION Plain olive, grey or brown above. Mainly smooth with orange- to yellowish-capped tubercles. Underside white with black flecking. **DISTRIBUTION** Found from Qld border, through central NSW, and into northern Vic. **HABITS AND HABITAT**

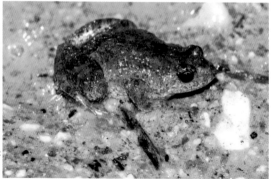

Inhabits woodland, swamps and agricultural areas, usually around slow-moving or still watercourses, including artificial dams and lakes. Feeds on invertebrates. Male's call, a short, mechanical chirp, is given while floating in a waterbody. Tadpoles start to develop into metamorphlings within 12 weeks of hatching.

South Coast Froglet ▪ *Crinia subinsignifera* TL 25mm

DESCRIPTION Pattern, colouration and texture of skin vary widely from brown, to light grey, to almost black above, with or without flecks and stripes. Back can be smooth, without bumps or protrusions, or extremely warty, and every variation in between. Underside white to black, with or with markings. **DISTRIBUTION** Found in south-west WA, from Manjimup to Cape Arid. **HABITS AND HABITAT** Lives in woodland with granite outcrops, around slow-moving or still watercourses, where it feeds on invertebrates. Call sounds like *wreeee*. Reproductive information is thought to be similar to that for the Bleating Froglet (see p. 66).

Tasmanian Froglet ▪ *Crinia tasmaniensis* TL 25mm

DESCRIPTION Colouration and texture of upperparts vary widely from yellow, to light grey, or almost black, with or without flecks and stripes. Back can be smooth, without bumps or protrusions, or extremely warty, and every variation in between. Underparts pink with black flecking. **DISTRIBUTION** Occurs in Tas. **HABITS AND HABITAT** Lives around slow-moving or still watercourses, including artificial dams, lakes and ornamental pools, where it feeds on invertebrates. Male's call, a sheep-like bleat, is repeated frequently, and is given from emergent vegetation, flooded grassland and pond edges. About 65 eggs are laid singly, sometimes attached to plant material, and tadpoles start to develop into metamorphlings 16 weeks after hatching.

Tinkling Froglet ▪ *Crinia tinnula* TL 18mm

DESCRIPTION Colouration and texture of skin vary widely from yellow, to light grey, or almost black above, with or without flecks and stripes. Back can be smooth, without bumps or protrusions, or extremely warty, and every variation in between. Underparts grey to brown, with a cream midline. **DISTRIBUTION** Found in coastal areas of south-east Qld,

from 1770 to the Illawarra region of NSW. **HABITS AND HABITAT** Lives around slow-moving or still watercourses, including wallum swamps, artificial dams, lakes and ornamental pools, where it feeds on invertebrates. Male's call is a *beeep*, repeated frequently, and is given while sitting on vegetation. About 70 eggs are laid singly, sometimes attached to plant material, and tadpoles start to develop into metamorphlings 10 weeks after hatching.

Smooth Frog ▪ *Geocrinia laevis* TL 35mm

DESCRIPTION Tan to dark brown or grey above with darker flecking. Some individuals have a paler triangle starting from in front of the eyes to the snout. Back relatively smooth and skin surface granular. Bluish-grey below, with dark speckling, and usually pinkish around groin. Throat of calling males generally yellowish. **DISTRIBUTION** Occurs in southwestern Vic and Tas. **HABITS AND HABITAT** Nocturnal. Lives in wet, closed

woodland and heaths, usually in moist pockets such as seepages and creek lines, and feeds on invertebrates. Call is a series of single squelches followed by a series of rapid clicking notes. About 150 eggs are laid, attached to vegetation in depressions, and they hatch when winter rains fill the depressions and the tadpoles enter the water. Tadpoles turn into frogs after about 25 weeks.

Lea's Frog ▪ *Geocrinia leai* TL 26mm

DESCRIPTION Tan to dark brown above, typically with pattern of broad, contrasting stripes, but can be plain. Some individuals have a broad stripe, starting from behind the eyes and extending down back, on to hips. Back relatively smooth and skin surface granular. Yellow to greenish below with dark speckling; throat of calling males usually

darker. **DISTRIBUTION** Occurs in southwestern WA, from west of Perth to Albany. **HABITS AND HABITAT** Nocturnal. Lives in wet, closed woodland and heaths, typically in moist pockets such as seepages and creek lines in the forest, where it feeds on invertebrates. Call is a series of single, *chirc*-like notes. About 70 eggs are laid attached to emergent vegetation, and the tadpoles fall into the surrounding water on hatching. They turn into frogs after about 16 weeks.

Victorian Smooth Frog
▪ *Geocrinia victoriana* TL 35mm

DESCRIPTION Tan to dark brown or grey above, and usually plain with darker flecking. Some individuals have a paler triangle starting from in front of the eyes to the snout. Back relatively smooth, and skin surface granular. Underside bluish-grey with dark speckling and pinkish around groin; throat of calling males usually yellowish.
DISTRIBUTION Found across Vic and into the border region of southern NSW.

With eggs

HABITS AND HABITAT Nocturnal. Lives in wet, closed woodland and heaths, usually in moist pockets such as seepages and creek lines in the forest, and feeds on invertebrates. Call is a series of single *raaak*, followed by series of rapid *pip pip pip pip* notes. About 120 eggs are laid, attached to vegetation in depressions. They hatch when winter rains fill the depression and the tadpoles enter the water. They turn into frogs after about 25 weeks.

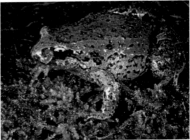

Forest Toadlet ■ *Metacrinia nichollsi* TL 25mm

DESCRIPTION Brown, slate-grey to black above, with flashes of bright colours visible from above in some individuals. Underparts grey, dark purple to black, with darker reticulations, and bright yellow, orange or red flash colours beneath each limb, and broader patch beneath hips. **DISTRIBUTION** Occurs in south-west WA, from Dunsborough to Albany. **HABITS AND HABITAT** Nocturnal. Lives in Karri and Jarrah woodland, where it is found beneath moist logs in damp depressions, and feeds on termites and ants. Burrows head-first into the soil. Male's call is a short *wrrk* repeated every few seconds, made from leaf litter. Lays about 20 eggs beneath logs, which hatch into small frogs after around 9 weeks.

Aaron Payne

Stuttering Frog ■ *Mixophyes balbus* TL 85mm

DESCRIPTION Tan to light brown or dark brown above, with black bands on legs, and white below. Back mainly smooth. Upper iris metallic blue. **DISTRIBUTION** Found from Tenterfield to Watagan National Park, NSW. Historically occurred into Vic. Probably a species complex.

HABITS AND HABITAT
Nocturnal. Lives around watercourses, where it feeds on invertebrates and small vertebrates. Male's call, a single *wha-rr-rup*, repeated frequently, is made from the edge of a waterbody, or occasionally from burrows or beneath leaf litter. Breeds in shallow riffles of permanent streams. About 600 eggs are laid into the water. Tadpoles start to develop into metamorphlings after 67 weeks.

Aaron Payne

Southern form

Scott Eipper

Carbine Barred Frog ■ *Mixophyes carbinensis* TL 78mm

DESCRIPTION Brown to grey-brown above, with broad darker brown longitudinal stripe along centre of back, and dark stripe from snout, extending over tympanum to shoulder. Back smooth without tubercles. White below. **DISTRIBUTION** Occurs in northeastern Qld, on Carbine and Windsor Tablelands. **HABITS AND HABITAT** Nocturnal. Found

in rainforests. Feeds on invertebrates and small vertebrates. Male's call, a double *wharrc*, is made from beneath leaf litter. Breeds in ephemeral pools beside streams. About 200 eggs are laid into the water and the female then kicks the fertilized eggs on to an overhanging bank above the water. When hatching, the tadpoles wriggle free and fall into the pool below. They start to develop into metamorphlings after a year.

Mottled Barred Frog ■ *Mixophyes coggeri* TL 105mm

DESCRIPTION Brown to grey-brown above, mottled with darker markings covering most of back, and with dark stripe from snout, extending over tympanum to shoulder. Back smooth, without tubercles. Underparts white. **DISTRIBUTION** Occurs in north-east Qld, from Cooktown to Paluma. **HABITS AND HABITAT** Nocturnal. Found in rainforests, where it feeds on invertebrates and small vertebrates. Male's call is a single

wharrrup, uttered from beneath leaf litter and repeated frequently. Breeds in ephemeral pools beside streams. About 500 eggs are laid into the water and the female then kicks the fertilized eggs on to an overhanging bank above the water. When hatching, the tadpoles wriggle free and fall into the pool below. They start to develop into metamorphlings after a year.

Great Barred Frog ■ *Mixophyes fasciolatus* TL 102mm

DESCRIPTION Tan or light brown to dark brown above, with black bands on legs. Back mainly smooth. Juveniles look similar to adults, but can have bright red or orange irises that fade to brown as they mature. Underparts white. **DISTRIBUTION** Occurs from eastern NSW to central-eastern Qld. **HABITS AND HABITAT** Nocturnal. Lives around watercourses in rainforests and wet sclerophyll forests, where it feeds on invertebrates and small vertebrates. Male's call is a single *wharrrup*, repeated frequently, and is given from

the edge of a waterbody, or occasionally from burrows or beneath leaf litter. Breeds in ephemeral pools beside streams. About 800 eggs are laid into the water and the female then kicks the fertilized eggs on to an overhanging bank above the water. When hatching, the tadpoles wriggle free and fall into the pool below. They start to develop into metamorphlings after a year.

Scott Eipper

Fleay's Barred Frog ■ *Mixophyes fleayi* TL 100mm

DESCRIPTION Tan or light brown to dark brown above, with black bands on legs. Back mainly smooth. Juveniles look similar to adults, but can have bright red or orange irises that fade to brown as they mature. Upper iris metallic green. Underparts white. **DISTRIBUTION** Found from Mt Warning, NSW, to Conondale ranges, Qld. **HABITS AND HABITAT** Nocturnal. Lives

around watercourses, where it feeds on invertebrates and small vertebrates. Male's call is a single *wharrrup*, repeated frequently, and is given from the edge of a waterbody, or occasionally from burrows or beneath leaf litter. Breeds in shallow sections of permanent streams. About 900 eggs are laid into the water, usually in a constructed depression among gravel, and tadpoles start to develop into metamorphlings after 30 weeks.

Scott Eipper

Giant Barred Frog ■ *Mixophyes iteratus* TL 120mm

DESCRIPTION Reddish-brown to dark brown above, with black bands on legs, and back mainly smooth. Underparts yellow. One of the most striking features of this species is its golden eyes. **DISTRIBUTION** Found from mid-eastern NSW, to southeastern Qld. **HABITS AND HABITAT** Nocturnal. Lives along watercourses, in rainforests and wet sclerophyll forests, where it feeds on invertebrates and small vertebrates. Male's call, a single *wharrrk*, is made from the edge of a waterbody, or occasionally from burrows or

beneath leaf litter. Breeds in ephemeral pools beside streams. Up to 4,200 eggs are laid into the water and the female then kicks the fertilized eggs on to an overhanging bank above the water. Eggs hatch 8–10 days later. When hatching, the tadpoles wriggle free and fall into the pool below, and they start to develop into metamorphlings after 40 weeks.

Northern Barred Frog ■ *Mixophyes schevilli* TL 93mm

DESCRIPTION Brown to reddish-brown above, with broad stripe of darker brown down centre line of back, and dark stripe from snout extending over tympanum to shoulder. Back smooth without tubercles. Underparts white. **DISTRIBUTION** Occurs in north-east Qld, on Cooktown and Kirrama Ranges. **HABITS AND HABITAT** Nocturnal. Found in

rainforests, where it feeds on invertebrates and small vertebrates. Male's call is a double *wharrc*, given from beneath leaf litter. Breeds in ephemeral pools beside streams. A record of 74 eggs laid is noted, but is thought to be unusually low. Tadpoles start to develop into metamorphlings after a year.

Turtle Frog ■ *Myobatrachus gouldii* TL 52mm

DESCRIPTION Typically yellow to brown above, although individuals can be reddish or slate-grey. Cream to white below. **DISTRIBUTION** Found in south-west WA, from Kalbarri to Esperance and Morowa. **HABITS AND HABITAT** Nocturnal. Lives on sandy soils in heaths, grassland and open woodland, where it feeds on termites and ants. Burrows head-first into the soil. Male's call, a *crraaawk*, is given from the surface or from inside the burrow. The attracted female shares the burrow with the male for up to three months. Lays about 35 eggs, which hatch into frogs after 7–9 weeks.

Angus McNab

Haswell's Frog ■ *Paracrinia haswelli* TL 39mm

DESCRIPTION Brown or tan to grey above, with black temporal stripe extending through tympanum and on to flanks, and thin pale vertebral stripe. Brightly coloured red flashes visible in groin. Upper iris usually pale to cream. Grey to black below, with white to cream spots and blotches. **DISTRIBUTION** Occurs in southeastern Australia, from Coffs Harbour, NSW, to Melbourne's eastern suburbs, Vic. **HABITS AND HABITAT** Nocturnal. Lives in forests, heaths and disturbed habitats, where it is typically found beneath moist logs and in vegetation. Feeds on invertebrates. Male's call is a short *eeep*, repeated every few seconds, and is given while floating in a waterbody. Lays about 190 eggs, which hatch into small frogs after around 9 weeks.

Scott Eipper

Red-crowned Broodfrog ■ *Pseudophryne australis* TL 31mm

DESCRIPTION Light grey to almost black above, with bright orange to red spots and flecks on top of head and along back and sides. Usually a red triangle from snout across top of head to rear of eyes. Underparts white with black marbling. **DISTRIBUTION** Found around Sydney, NSW. **HABITS AND HABITAT** Nocturnal. Lives in moist gutters and seepages, formed between layers of sandstone where water gathers and is held after rain,

and feeds mainly on ants. Male's call is a short squelch, which is given from beneath leaf litter or from under rocks. About 45 eggs are laid in a clump in leaf litter or soil, and are guarded by the male until the nest is inundated by rain, following which the eggs hatch into tadpoles that swim into the new pool. Metamorphosis takes about 14 weeks.

Bibron's Broodfrog ■ *Pseudophryne bibronii* TL 31mm

DESCRIPTION Light grey to almost black above, with bright yellow, orange or red flash colours in armpits, groin and on the urostyle. Underparts white with black mottling. **DISTRIBUTION** Found from south-east Qld, across southeastern Australia, to southern SA. Possibly more than one species. **HABITS AND HABITAT** Nocturnal. Lives in moist dry creek lines and seepages in open woodland, heaths, grassland and swamps, where it

feeds mainly on ants. Male's call, a short squelch, is made from beneath leaf litter or vegetation, or from under rocks. About 45 eggs are laid in a clump in leaf litter or soil, which are guarded by the male until the nest is inundated by rain, after which the eggs hatch into tadpoles that swim into the new pool. Metamorphosis takes about 20 weeks.

Red-backed Broodfrog ■ *Pseudophryne coriacea* TL 25mm

DESCRIPTION Light tan to dark reddish-brown above, with varying degrees of brightness on back, and often sparsely flecked with black. Flanks strongly delineated, being grey to black. White below with black mottling. **DISTRIBUTION** Found from the south-east Qld border, south to Watagan National Park, NSW. **HABITS AND HABITAT** Nocturnal. Lives in moist dry creek lines and seepages within woodland, where it feeds mainly on

ants. Male's call, a short squelch, is made from beneath leaf litter or vegetation, or from under rocks. About 70 eggs are laid in a clump within leaf litter or soil, which are guarded by the male until the nest is inundated by rain, after which the eggs hatch into tadpoles that swim into the new pool. Metamorphosis takes about 8 weeks.

Scott Eipper

Southern Corroboree Frog ■ *Pseudophryne corroboree* TL 29mm

DESCRIPTION Top of body striped in black and yellow, and ventral surface black and white. **DISTRIBUTION** Restricted to single site in Kosciuszko National Park, NSW, but formerly of a larger area in the Australian Alps. Range greatly contracted due to chytrid fungus and habitat destruction. **HABITS AND HABITAT** Lives in sphagnum bogs, where it mainly feeds on small invertebrates such as ants. About 25 eggs are laid above the water,

which are guarded by the adults until water levels increase in the depression, inundating the clutch and causing the eggs to hatch. Tadpoles take about 6–8 months to complete metamorphosis. One of only a few Australian frog species that produce alkaloid toxins similar to those of the poison dart frogs of South America. Alkaloid toxins are also manufactured from its food.

Scott Eipper

Magnificent Broodfrog ■ *Pseudophryne covacevichae* TL 28mm

DESCRIPTION Reddish-brown to grey above, with yellow bands on arms and on urostyle. White below with black mottling. **DISTRIBUTION** Found in north-east Qld, from Paluma to Ravenshoe. **HABITS AND HABITAT** Nocturnal. Lives in moist dry creek

lines and seepages in open woodland, where it feeds mainly on ants. Male's call, a short squelch, is made from below leaf litter or vegetation, or from under rocks. About 50 eggs are laid within leaf litter or soil, which are guarded by the male until the nest is inundated by rain, after which the eggs hatch into tadpoles that swim into the new pool. Metamorphosis takes about 7 weeks.

Dendy's Broodfrog ■ *Pseudophryne dendyi* TL 29mm

DESCRIPTION Light grey to almost black above, with bright yellow markings on crown, arms, legs, groin and urostyle. Some stunning individuals are almost entirely yellow. Underparts white with black mottling. **DISTRIBUTION** Found from eastern Vic, north-east into NSW to about Jervis Bay inland to the ACT. **HABITS AND HABITAT** Nocturnal. Known to hybridize with Bibron's Broodfrog (see p. 78) where ranges overlap, and possibly a colour variant of it as currently defined. Found along moist dry creek

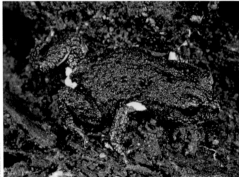

lines and seepages in open woodland, heaths, grassland and swamps, where it feeds mainly on ants. Male's call, a short squelch, is made from below leaf litter or vegetation, or from under rocks. About 70 eggs are laid in a clump in leaf litter or soil, which are guarded by the male until the nest is inundated by rain, after which the eggs hatch into tadpoles that swim into the new pool. Metamorphosis takes around 30 weeks.

Gorge Broodfrog ■ *Pseudophryne douglasi* TL 32mm

DESCRIPTION Light grey to dark reddish-brown above, mottled with darker blotches or spots, and normally with low tubercles. Sometimes has orange to red blotches, mainly on head, urostyle and limbs. Grey below with black mottling. **DISTRIBUTION** Found in central WA, from Cape Range to Millstream. **HABITS AND HABITAT** Nocturnal. Lives in moist creek lines and gorges, where it feeds mainly on ants. Male's call is similar to a sharp *weeek*, and is made from within burrows, below vegetation or under rocks. About 90 eggs are laid in a clump beneath stones in shallow water, and metamorphosis takes about 20 weeks.

Brad Maryan

Crawling Broodfrog ■ *Pseudophryne guentheri* TL 30mm

DESCRIPTION Whitish-grey to brown above with darker mottling, and occasionally with pale triangle on crown. Covered in bumps and low tubercles, which can be capped with yellow or red. Underparts white with limited black mottling. **DISTRIBUTION** Found across south-west WA, from Shark Bay to Israelite Bay. **HABITS AND HABITAT** Nocturnal. Lives in moist dry creek lines and seepages in open woodland, granite hills, swamps and heaths, where it feeds mainly on ants. Male's call, a short squelch, is made from below leaf litter or vegetation, or from under rocks. About 120 eggs are laid in a clump within breeding tunnels, and metamorphosis takes about 8 weeks.

Grant Webster

Large Broodfrog ▪ *Pseudophryne major* TL 31mm

DESCRIPTION Light grey to almost black or reddish-brown above, with bright yellow flash colours in armpits and groin, and on urostyle. Lower flanks can have fine pale blue flecks. Underparts white with black mottling. **DISTRIBUTION** Found from northern Gold Coast to about Bowen, Qld. Possibly more than one species. **HABITS AND HABITAT** Nocturnal. Lives in moist dry creek lines and seepages in woodland and swamps, where

it feeds mainly on ants. Male's call, a short squelch, is made from below leaf litter or vegetation, or from under rocks. One clutch of 112 eggs has been noted. Eggs are laid in a clump in leaf litter or soil, and are guarded by the male until the nest is inundated by rain, after which the eggs hatch into tadpoles that swim into the new pool. Metamorphosis takes about 20 weeks.

Inland form

Western Broodfrog ▪ *Pseudophryne occidentalis* TL 31mm

DESCRIPTION Light grey to dark reddish-brown above, mottled with darker blotches or spots, and usually with an orange triangular marking on crown, tops of arms and urostyle. Typically covered in low, flattened tubercles. Underparts white with black mottling.

DISTRIBUTION Found in WA, from Shark Bay through inland areas to Israelite Bay. **HABITS AND HABITAT** Nocturnal. Lives in moist areas on granite outcrops and surrounding low-lying areas, where it feeds mainly on ants. Male's call is similar to a sharp *weeek*, and is made from burrows or under rocks. Around 90 eggs are laid in a clump in breeding tunnels, and metamorphosis takes about 8 weeks.

Anders Zimny

Northern Corroboree Frog ▪ *Pseudophryne pengilleyi* TL 26mm

DESCRIPTION Top of body striped in black and greenish-yellow. Ventral surface black and white. **DISTRIBUTION** Known from a few sites in Fiery and Brindabella Ranges of NSW and ACT. Formerly of a larger area, the range greatly contracted due to chytrid fungus and habitat destruction. **HABITS AND HABITAT** Lives in sphagnum bogs, where it mainly feeds on small invertebrates such as ants. This habitat has become endangered due to damage from horses, drought and fire. About 25 eggs are laid above the water, which are guarded

by the adults until water levels increase in the depression, inundating the clutch and causing the eggs to hatch. More than one female may lay eggs in the same nest site. Tadpoles take about 30 weeks to complete metamorphosis. One of only a few Australian frog species that produce alkaloid toxins similar to those of the poison dart frogs of South America. Alkaloid toxins are also manufactured from its food.

Aaron Payne

Copper-backed Broodfrog ■ *Pseudophryne raveni* TL 25mm

DESCRIPTION Light tan to dark reddish-brown above, with varying degrees of brightness on back, and often sparsely flecked with black. Flanks strongly delineated, being light grey to black. Underparts are white with black mottling. **DISTRIBUTION** Found in Qld, from Rockhampton to far northern NSW. Isolated population at Eungella, Qld. **HABITS AND HABITAT** Nocturnal. Lives in moist dry creek lines and seepages in woodland, where it feeds mainly on ants. Male's call, a short squelch, is made from below leaf litter or vegetation, or from under rocks. One breeding record of 56 eggs, laid in a clump within leaf litter or soil. Eggs are guarded by the male until the nest is inundated by rain, after which they hatch into tadpoles that swim into the new pool. Metamorphosis is thought to take about 20 weeks.

Scott Eipper

Central Ranges Broodfrog ▪ *Pseudophryne robinsoni* TL 30mm

DESCRIPTION Light grey to dark brown above, mottled with darker blotches or spots, and typically with orange markings on tops of arms, legs and urostyle. Usually covered in low, flattened tubercles. Black below with white mottling. **DISTRIBUTION** Found in SA, in Everard and Musgrave Ranges. **HABITS AND HABITAT** Male's call is similar to a sharp *eeeep*, and is made from burrows or under rocks. Reproductive biology thought to be similar to that of the Western Broodfrog (see p. 83).

Aaron Payne

Southern Broodfrog ▪ *Pseudophryne semimarmorata* TL 32mm

DESCRIPTION Grey to blue above, but occasionally metallic green with darker spots and blotches, and covered in small tubercles. Ventral surface orange or yellow with black and white. **DISTRIBUTION** Found in eastern Tas and southern Vic. **HABITS AND HABITAT** Lives in grassland, heaths and forests in low-lying depressions, where it feeds on ants.

Has significantly declined due to chytrid fungus and habitat destruction. Lays about 130 eggs in burrows constructed beneath vegetation in depressions that fill with autumn rains. When the water levels increase, the depression is inundated, causing the eggs to hatch. Tadpoles take around 30 weeks to complete development.

Scott Eipper

Southern Gastric-brooding Frog ■ *Rheobatrachus silus* TL 52mm

DESCRIPTION Brown to blackish-grey above; some individuals mottled with darker markings, and covered in low tubercles. Underside yellow to orange beneath legs.
DISTRIBUTION Historically found from Blackall Ranges to the Conondales, Qld. Last

wild individuals seen at the Conondales in 1981. Thought to have become extinct due to chytrid fungus. **HABITS AND HABITAT** Diurnal. Formerly found along rainforest streams and creeks, sheltering beneath stones in streams and feeding on invertebrates. Call a slow *wrrraack*. Laid about 30 eggs, which were ingested by the female which, while brooding young, stopped producing digestive acids. Young emerged as fully formed frogs 6 weeks later.

Northern Gastric-brooding Frog ■ *Rheobatrachus vitellinus* TL 79mm

DESCRIPTION Brown to blackish-grey above; some individuals mottled with darker markings, and covered in low tubercles. Underside yellow. **DISTRIBUTION** Historically found around Eungella, Qld. Last wild individuals seen at Finch Hatton, Qld, in 1986.

Thought to have become extinct due to chytrid fungus. **HABITS AND HABITAT** Diurnal. Formerly found along rainforest streams and creeks, sheltering beneath stones in streams and feeding on invertebrates. Call a slow *wrrraaaaack*. Laid about 22 eggs, which were ingested by the female which, while brooding young, stopped producing digestive acids.

Sunset Frog ▪ *Spicospina flammocaerulea* TL 36mm

DESCRIPTION Dark blue-black above, covered in low bumps, and with orange feet and hands. Large, elongated parotoid glands. Underparts bright orange and blue, with darker marbling through the blue.

DISTRIBUTION Found in south-west WA, near Walpole. **HABITS AND HABITAT** Nocturnal, Lives in peat swamps and bogs, where it feeds on invertebrates. Male's call is a two-part *bah-buck*, and is given from open water or beneath floating vegetation. Lays about 100 eggs in algae mats, which hatch into tadpoles that take about 11 weeks to become frogs.

Grant Webster

Sharp-nosed Day Frog ▪ *Taudactylus acutirostris* TL 29mm

DESCRIPTION Reddish to yellowish-brown on the back, becoming dark grey to charcoal on flanks, and with thin, white to cream dorsolateral stripe extending from snout to groin. Rear legs banded. Some individuals had dark markings between shoulders. Underparts grey with white to yellow spotting and flecks. Upper jaw extended much further than lower jaw, elongating the snout. **DISTRIBUTION** Historically found from Big Tableland to Cardwell, Qld, at 300–1,300m

asl. Last wild individual seen at Mt Hartley in 1997. Thought to have become extinct due to chytrid fungus. **HABITS AND HABITAT** Diurnal. Was found along rainforest streams and creeks, where it fed on invertebrates. Male's call consisted of a series of short, metallic chirps and clicks, made from stream-side rocks and leaf litter. Laid about 30 eggs beneath rocks.

Hal Cogger

Southern Day Frog ■ *Taudactylus diurnus* TL 30mm

DESCRIPTION Yellowish-brown to grey above, mottled with darker markings, and with dark marking between shoulders in some individuals. Underparts grey with white to yellow spotting. **DISTRIBUTION** Historically found from Blackall Ranges to Mt Glorious, Qld.

Last wild individuals seen at Green's Falls in 1979. One of the first species in Australia to disappear due to chytrid fungus. **HABITS AND HABITAT** Diurnal. Was found along rainforest streams and creeks and fed on invertebrates. Male's call consisted of a series of short clucks, made from stream-side rocks and leaf litter. Laid about 30 eggs beneath rocks.

Hal Cogger

Eungella Day Frog ■ *Taudactylus eungellensis* TL 36mm

DESCRIPTION Yellowish-brown to grey above, mottled with darker markings, and some individuals have dark, 'X'-shaped marking between the shoulders. Back covered in scattered tubercles. Young individuals and juveniles occasionally marbled with lime green. Underparts yellow to white, occasionally with flecking. **DISTRIBUTION** Found in mid-eastern Qld around Eungella National Park, but has suffered significant declines due

to chytrid fungus. **HABITS AND HABITAT** Cathemeral. Found along rainforest streams and creeks, and usually seen in splash zones of fast-moving streams, where it feeds on invertebrates. Male's call, a series of short, metallic clicks, is made from stream-side rocks while sitting on fallen palm fronds, but also uses arm and leg waving to communicate with other frogs. Lays about 45 eggs beneath rocks.

Scott Eipper

Eungella Tinker Frog ■ *Taudactylus liemi* TL 30mm

DESCRIPTION Orange-brown, reddish to grey above, mottled with darker markings. Some individuals have dark, 'X'-shaped marking between shoulders. Back smooth with a few scattered tubercles. Flanks can be the same colour as back or sharply delineated with dark grey to charcoal.

Underparts yellow to white, occasionally with flecking. **DISTRIBUTION** Found in mid-eastern Qld around Eungella National Park, but has suffered declines due to chytrid fungus. **HABITS AND HABITAT** Cathemeral. Occurs along rainforest soaks and weeps that drain into streams, where it feeds on invertebrates. Male's call consists of 4–8 short, metallic tinks, and is given from crevices and beneath rocks. Lays about 45 eggs beneath rocks.

Angus McNab

Kroombit Tinker Frog ■ *Taudactylus pleione* TL 30mm

DESCRIPTION Brown to grey-brown above, mottled with darker markings, and usually dotted with white and cream. Most individuals have dark, 'X'-shaped marking between the shoulders. Back smooth, with a few scattered tubercles. Flanks delineated with dark grey to charcoal. Underparts grey with flecking. **DISTRIBUTION** Found in mid-eastern Qld, around Kroombit Tops, but has suffered significant declines due to chytrid fungus, and because of very restricted range is at serious risk of extinction.

HABITS AND HABITAT Crepuscular to nocturnal. Found in rocky rainforest gullies dominated by palms, where it feeds on invertebrates. Male's call is made up of 6–11 short, metallic tinks, and is given from crevices and beneath rocks. Males are very selective on call location, using the surrounding structure to naturally amplify their call. Reproductive biology unknown.

Scott Eipper

Northern Tinker Frog ■ *Taudactylus rheophilus* TL 30mm

DESCRIPTION Reddish to yellowish-brown on back, with extensive mottling in some individuals. Flanks dark blue-grey to charcoal, and rear legs banded. Underparts are brown with yellow spotting and flecks. **DISTRIBUTION** Historically found from Thornton

Peak to Mt Bellenden Ker, Qld, at 700–1,300m asl. Last wild specimen seen at Mt Bellenden Ker in 2000. Probably extinct due to chytrid fungus. **HABITS AND HABITAT** Cathemeral. Was found along rainforest streams and creeks, where it fed on invertebrates. Male's call, which consisted of 6–11 short, metallic tinks, was given from rocks and vines along creeks. Laid about 40 eggs beneath rocks.

Montane Toadlet ■ *Uperoleia altissima* TL 30mm

DESCRIPTION Generally pale grey to brownish above, indiscriminately mottled blackish and spotted with pale creamish-yellow. Black markings form a rearward 'V' between eyes. Skin covered with numerous rounded tubercles. Underparts creamish with darker flecking, and with bright red patch in groin and inner thigh (behind knee). Toes unwebbed, and females larger than males. **DISTRIBUTION** Found in higher altitudes of Atherton and

Windsor Tablelands of far northeastern Qld, typically on western side. **HABITS AND HABITAT** Inhabits wetlands, swamps, moist sclerophyll forests and woodland, where it feeds mainly on invertebrates, and is mostly active during the wet season. Call is as a short, squelchy *clik*, which distinguishes it from other similar frogs within its range. Lays about 120 eggs in still pools, and tadpoles take about 5 weeks to complete metamorphosis.

Jabiru Toadlet ■ *Uperoleia arenicola* TL 23mm

DESCRIPTION Dark grey to brown above, with both dark and light mottling, and rear of thighs reddish-orange. White to cream below, with throat grey in calling males. Very similar to Fat Toadlet (see p. 92). **DISTRIBUTION** Restricted to small area in western Arnhem Land, near Jabiru, Alligator Rivers, northwestern NT. **HABITS AND HABITAT** Occupies fringes of creeklines with sandy soils at bases of sandstone escarpments, where it burrows underground, emerging during wet season to breed. Call is in two parts, sounding similar to *wirr-wrink*. Reproductive biology largely unknown.

Aaron Payne

Derby Toadlet ■ *Uperoleia aspera* TL 35mm

DESCRIPTION Dark grey to brown above, with mixture of dark and light mottling, and covered in numerous raised tubercles. Hindlimbs short, toes with only minor webbing and thighs reddish-orange on lateral edges. White to cream below, with throat grey in calling males. **DISTRIBUTION** Occurs in a small region of western Kimberley, WA, from Dampier Peninsula to Fitzroy Crossing. **HABITS AND HABITAT** Nocturnal. Inhabits seasonally wet grassland bordering on rocky gravel areas, where it is restricted to rain-filled depressions and adjacent grassy tussocks. Call consists of repeated loud clicks. Lays about 120 eggs in still pools, and tadpoles take about 7 weeks to complete metamorphosis.

Aaron Payne

Northern Toadlet ■ *Uperoleia borealis* TL 32mm

DESCRIPTION Generally dull brown to dark grey above and paler below, with grey stippling on throat. Hindlimbs short, with bright orange patches on groin and rear of thighs, and toes have moderate webbing. **DISTRIBUTION** Found across Kimberley and

Ord regions of WA, and northwestern NT. **HABITS AND HABITAT** Nocturnal. Inhabits open grassland and rocky creek lines that are subject to temporary flooding during wet season. During breeding season occupies grassy tussocks or shelters among rocky areas. Males emit a short, rasping call. Lays about 320 eggs in still pools. Tadpole development time unknown.

Fat Toadlet ■ *Uperoleia crassa* TL 34mm

DESCRIPTION Generally dull brown to dark grey above, variably patterned, from immaculate to heavily mottled. Body covered in low tubercles. Paler below, with grey stippling on throat. Hindlimbs short, with bright orange patches on groin and rear of thighs, and toes have moderate webbing. This species was recently shown to be the same as **Floodplain Toadlet** *U.inundata*. **DISTRIBUTION** Found in northern Australia

from Mitchell Plateau, WA, across northen NT and adjacent parts northwestern Qld. **HABITS AND HABITAT** Nocturnal. Inhabits open areas in open forests that are temporarily flooded during wet season, and shelters in bases of grassy tussocks during this time. Males emit a rasping call to attract a mate. Lays about 300 eggs in still pools. Tadpole development time unknown.

Howard Springs Toadlet ■ *Uperoleia daviesae* TL 22mm

DESCRIPTION Generally dull brown to dark grey above with a variable pattern, from immaculate to heavily mottled, and often with dull reddish vertebral stripe. Body covered in low tubercles. Hindlimbs short, with bright orange patches on groin and rear of thighs, and toes have moderate webbing. Paler below, with grey stippling on throat. **DISTRIBUTION** Restricted to Howard River and Elizabeth River catchments near Humpty Doo, NT. **HABITS AND HABITAT** Nocturnal. Occupies patches of flooding sandy soils in grassy woodland,

particularly where carnivorous plants are common, where it spends drier part of the year sheltering under fallen timber or other ground debris, emerging after heavy rains to congregate around shallow pools. Males emit a rasping call to attract a mate, from bases of soil rises and termite mounds. Reproductive biology is unknown. Currently classed as Vulnerable in NT but is being reviewed.

Ryan Francis

Dusky Toadlet ■ *Uperoleia fusca* 30mm

DESCRIPTION Generally grey to brown above, blotched darker, and with numerous scattered short tubercles, occasionally tipped with orange (more so in younger individuals). Legs have darker barring, and groin and rear of knees have yellow to red patches. Toes without webbing. Underparts pale, strongly peppered with dark grey or dark brown. Female typically larger than male.

DISTRIBUTION Found along GDR from Gladstone, Qld, to around Kurrajong NSW; isolated population near Eungella, Qld. **HABITS AND HABITAT** Nocturnal. Inhabits forests and grassy woodland, where it breeds in rain-filled, grassy depressions and dams. Call consists of loud, penetrating, raspy clicks. Lays about 300 eggs in still pools, and tadpoles take about 12 weeks to complete metamorphosis.

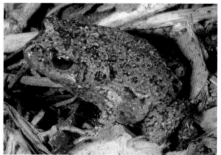

Scott Eipper

Glandular Toadlet ▪ *Uperoleia glandulosa* TL 30mm

DESCRIPTION Generally dull brown to dark grey above with variable pattern, from immaculate to heavily mottled, and often with dull reddish vertebral stripes. Body covered in low tubercles. Paler below, with grey stippling on throat. Hindlimbs short, with bright orange patches on groin and rear of thighs, toes have moderate webbing. DISTRIBUTION Restricted to Roeburn Plain and Yule River Valley of Pilbara, WA. HABITS AND HABITAT Nocturnal. Lives in spinifex-dominant floodplains, emerging after heavy rains to feed and breed. Male's call is a short *cric*, given from bases of soil rises and termite mounds. Reproductive biology poorly known.

Aaron Payne

Gurrumul's Toadlet ▪ *Uperoleia gurrumuli* TL 31mm

DESCRIPTION Colouration based on preserved specimens. Upperparts generally uniform grey, reddish to grey-brown, with irregular shadings and mottling of blackish-brown. Covered in low tubercles. Underside white to grey, and throat of calling males dark grey. DISTRIBUTION Found on Wessel Islands (Djirrkari, Guluwurru, Martjanba and

Renee Catullo

Rarrakala). HABITS AND HABITAT Most similar to the Fat Toadlet (see p. 92). Nocturnal. Habitat unknown but probably found in open woodland, where it feeds on invertebrates. Reproductive biology and male call currently unknown, but probably a 'squelching' sound based on its similarity to the Fat Toadlet.

Smooth Toadlet ■ *Uperoleia laevigata* TL 29mm

DESCRIPTION Generally grey to brown above, clearly blotched with darker patches, and with large parotoid glands. Covered with numerous scattered short tubercles, occasionally tipped with orange. Legs have darker barring, and groin and rear of knees have red patches; toes without webbing. Underparts grey, strongly peppered with dark grey or dark brown. **DISTRIBUTION** Found along coast and ranges of eastern Australia from Cann River to Newcastle. Occurs on western

slopes of GDR and Lockyer Valley to Rockhampton Qld. **HABITS AND HABITAT** Nocturnal. Inhabits dry forests and grassy woodland, where it breeds in rain-filled, grassy depressions and dams. Call consists of loud, penetrating, raspy clicks. Lays about 300 eggs in still pools, and tadpoles take about 16 weeks to complete metamorphosis.

Scott Eipper

Stonemason Toadlet ■ *Uperoleia lithomoda* TL 32mm

DESCRIPTION Dark olive-green to brownish above, with darker blotches and scattered paler spots. Skin smooth, with numerous conspicuous raised tubercles and prominent parotoid glands. Underparts creamish, with a blackish line following the edge of lower jaw. Females larger than males. **DISTRIBUTION** Widespread throughout northern Australia, from Ord region, WA, through northern NT and Cape York Peninsula, to central eastern Qld. **HABITS AND**

HABITAT Nocturnal. Inhabits woodland and grassland, where it is most commonly encountered during the wet season, when numbers congregate around rain-filled, grassy hollows for breeding. Feeds mainly on invertebrates. Call is an abrupt, explosive *tok*. Lays about 320 eggs in still pools, and tadpoles take about 12 weeks to complete metamorphosis.

Scott Eipper

Littlejohn's Toadlet ■ *Uperoleia littlejohni* TL 29mm

DESCRIPTION Greyish above, with large, variable dark olive-brown blotches edged with blackish and smaller blackish flecks. The numerous raised tubercles are tipped with orange, and parotoid glands are prominent. Underparts greyish-white with darker grey

to brown peppering. Throat black on calling males. Toes partially webbed at bases. **DISTRIBUTION** Found in central northeastern Qld. **HABITS AND HABITAT** Nocturnal. Inhabits open forests, woodland, rocky gorges and grassland, where it is most commonly encountered during the wet season, when numbers congregate around rain-filled depressions for breeding. Feeds mainly on invertebrates. Call is an abrupt, explosive *tok*.

Mahony's Toadlet ■ *Uperoleia mahonyi* TL 32mm

DESCRIPTION Generally grey to brown above, clearly blotched with darker patches, and lower flanks can be whitish. Covered with numerous scattered, short tubercles, occasionally tipped with orange. Legs have darker barring, and groin and rear of thighs have orange patches. Underparts grey with black marbling and white spots. **DISTRIBUTION** Currently only known from coastal swamps on Central Coast region of NSW. **HABITS AND HABITAT** Nocturnal. Recently described. Inhabits coastal swamps. Call consists of a single *wraa*, repeated about 1.5 seconds apart. Tadpoles raised in a laboratory took 58 days to complete metamorphosis. Classed as Endangered in NSW.

Marbled Toadlet ▪ *Uperoleia marmorata* TL 30mm

DESCRIPTION Known from a single specimen collected in 1841. The specimen was described in life by the collector as being marbled black and green above, with a triangular green spot on the forehead. Has maxillary teeth and a trace of webbing between toes. The only other species in the Kimberley with teeth is Mjoberg's Toadlet (see p. 100), which has a distinct tubercle on the heel that is not noted on the type specimen of the Marbled Toadlet. **DISTRIBUTION** The single specimen is thought to have come from near the mouth of the Prince Regent River in the Kimberley region of WA. **HABITS AND HABITAT** Details of natural history unknown.

Martin's Toadlet ▪ *Uperoleia martini* TL 37mm

DESCRIPTION Generally grey to brown above with darker patches, and usually with paler triangular crown between eyes and nostrils. Covered with numerous short tubercles, occasionally tipped with orange, yellow or brown, and has large, prominent parotoid glands. Upper arms bright yellow. Legs have darker barring, and groin and rear of thighs have yellow to orange-red patches. Underparts dark grey with black marbling and white spots. Very similar to Tyler's Toadlet (see p. 104) but differs by the call, which is longer, and colouration of underside (bluish-grey in Tyler's Toadlet). **DISTRIBUTION** Occurs from Eden, NSW, along coast, through Gippsland, to Rosedale, Vic. **HABITS AND HABITAT** Nocturnal. Inhabits swamps and dams in coastal dunes, heaths and closed woodland. Male's call, a single *wraaaa* repeated about 2 seconds apart, is made from beneath grass tussocks and sedges near the water's edge. Tadpoles take about 18 weeks to complete metamorphosis. Regarded as Endangered in Vic.

Tiny Toadlet ■ *Uperoleia micra* TL 22mm

Paul Doughty/WA Museum

DESCRIPTION Dark brown to blackish above with darker mottled spotting, and covered in large tubercles. Pale blue speckling on lower flanks. Ventral surface white to cream with grey stippling. **DISTRIBUTION** Found at Prince Regent River, Walcott Inlet and a couple of islands in Kimberley region, WA. **HABITS AND HABITAT** Nocturnal. Inhabits sandstone rocky outcrops in savannah with permanent water, such as rock pools. During breeding periods males emit a short *eek*. Reproductive biology poorly known.

Tanami Toadlet ■ *Uperoleia micromeles* TL 36mm

DESCRIPTION Generally dull brown to pale grey above and paler below, with grey stippling on throat. Variably patterned, but often heavily mottled, and with dull reddish vertebral stripe. Body covered in low red tubercles. Hindlimbs short and toes have moderate webbing; bright orange patches on groin and rear of thighs. **DISTRIBUTION** Only found in Tanami and Great Sandy Deserts of western NT, across to Port Hedland, WA. **HABITS AND HABITAT** Nocturnal. Lives on red sand with Mulga and spinifex, emerging after heavy rains to feed and breed. Shelters beneath rocks and soil. Call is unknown. Reproductive biology poorly known.

Hal Cogger

Mimic Toadlet ■ *Uperoleia mimula* TL 32mm

DESCRIPTION Greyish to brown above, with variable darker brown mottling. Upperparts have scattered raised tubercles and parotoid glands are prominent and blotched. Underparts greyish-white with darker grey to brown peppering, and groin has conspicuous orange-red patch. Toes partially webbed at bases. **DISTRIBUTION** Found in northeastern Qld and islands of Torres Strait. **HABITS AND HABITAT** Inhabits open woodland and grassland, where it is active during the wet season, at which time numbers congregate around rain-filled depressions for breeding. Call is a regularly repeated, clicking *riik*. Reproductive biology poorly known.

Anders Zimny

Small Toadlet ■ *Uperoleia minima* TL 23mm

DESCRIPTION Dark brown to blackish above with darker mottled spotting, and covered in large yellow to orange tubercles. Ventral surface grey with white stippling, and rear of thighs has bright red patches. **DISTRIBUTION** Found across Kimberley Plateau region, WA. **HABITS AND HABITAT** Nocturnal. Inhabits savannah and grassland in low-lying areas prone to flooding. During breeding periods the males emit a short click. Reproductive biology poorly known.

Marion Anstis

Mjoberg's Toadlet ■ *Uperoleia mjobergii* TL 25mm

DESCRIPTION Grey to deep creamy-brown above, with darker blotches and yellowish to reddish patches, most noticeable on parotoid glands. Hindlimbs short and obscurely barred.

Bright orange-red patches on groin and rear of thighs. DISTRIBUTION Found in western Kimberley, WA, between Fitzroy Crossing and Dampier Peninsula. HABITS AND HABITAT Nocturnal. Inhabits claypans, savannah and grassland, mainly in low-lying areas prone to flooding. During breeding periods the males emit a short *riiick*. Reproductive biology poorly known.

Alexandria Toadlet ■ *Uperoleia orientalis* TL 28mm

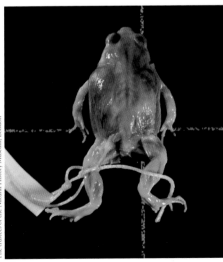

DESCRIPTION Very poorly known. Thought to be grey to light brown. Two of the three individuals recorded had a pale vertebral stripe and distinct reddish parotoid glands. Body covered in low tubercles. Underside white to cream, and throat brown in males. Hindlimbs short and possibly banded from above. Toes webbed. DISTRIBUTION Known from three individuals in NT; two were found on Alexandria Station, however the size of this station was considerably larger at the time of discovery compared with today; a third specimen found at Arnold River in 1977 in a paperbark swamp fed by a spring. HABITS AND HABITAT Essentially unknown.

Wrinkled Toadlet ■ *Uperoleia rugosa* TL 32mm

DESCRIPTION Grey-brown above with irregular darker blotches and paler cream patches (especially on parotoid glands), and orange spotting. Head small with darker triangular patch. Skin smooth, often with numerous raised tubercles, giving rough appearance. Underparts whitish, becoming more greyish on throat, and with some brownish speckling. Groin has bright scarlet patches, and toes are unwebbed with narrow to broad fringes.
DISTRIBUTION Occurs in southern Qld, central and western NSW, northern Vic and far eastern SA. **HABITS AND**

HABITAT Nocturnal. Found in dry forests, woodland and grassland subjected to flooding after heavy rains, where it congregates around waterholes, streams and temporarily flooded areas to breed. Burrows underground at other times. The Small-headed Toadlet *U. capitulata* now merged with this species. Call is a short *crark*. Lays about 200 eggs in still pools, and tadpoles take about 7 weeks to complete development.

Aaron Payne

Russell's Toadlet ■ *Uperoleia russelli* TL 35mm

DESCRIPTION Generally grey above, often with heavy mottling. Parotoid glands, along with many of the numerous tubercles, pale orange to brown. There may be a vertebral stripe of orange-capped tubercles. Underside grey with pink limbs and small red patches on rear of thighs. Toes

have extensive webbing.
DISTRIBUTION Occurs in WA, from Shark Bay to near Exmouth, including Gascoyne and Carnarvon regions.
HABITS AND HABITAT Nocturnal. Found along flowing watercourses and surrounding open woodland. Males call from above the water among leaves and other undergrowth, emitting a short *warrk*. Reproductive biology poorly known.

Marion Anstis

Pilbara Toadlet ■ *Uperoleia saxatilis* TL 37mm

DESCRIPTION Generally brown above, often with heavy mottling. Usually has red to orange submandibular glands. Numerous tubercles dark brown. May also have a vertebral stripe of orange-capped tubercles. Underside grey with limbs being pink. Toes lack extensive webbing. Small red patches on rear of thighs. **DISTRIBUTION** Found in Pilbara

region, WA. **HABITS AND HABITAT** Nocturnal. Occurs along watercourses in rocky areas. Previously thought to be synonymous with Russell's Toadlet (see p. 101); difference in distribution is the easiest way to distinguish the species. Males call from above the water beneath spinifex and in rock crevices. Male's call is a short *warrk*. Reproductive biology poorly known.

Ratcheting Toadlet ■ *Uperoleia stridera* TL 25mm

DESCRIPTION Body dorsolaterally compressed. Orange-brown to dark grey above, often with dull reddish vertebral stripe, and covered in orange, conical tubercles. Underside white with grey flecking, and throat darker in breeding males. Hindlimbs short and toes without webbing, and bright orange-red patches on groin and rear of thighs.

DISTRIBUTION Found from Daly Waters region, NT, to Fitzroy Crossing, WA. **HABITS AND HABITAT** Nocturnal. Lives in grassy floodplains and claypans, usually on cracking black soil, emerging after heavy rains to feed and breed. Male's call is a short *cric*, repeated every 0.5 of a second. Reproductive biology poorly known.

Mole Toadlet ■ *Uperoleia talpa* TL 40mm

DESCRIPTION Bronzed brown above with irregular blackish markings and some reddish spots. Skin smooth with some scattered, raised tubercles. Underparts creamish-white and throat black, forming rim around lower jaw, and underside of limbs purple. Toes moderately webbed and broadly fringed. Female typically larger than male. **DISTRIBUTION** Found in Derby region, western Kimberley, WA. **HABITS AND HABITAT** Inhabits sparsely vegetated sandy plains, where it breeds during the wet season (mainly February) in shallow, water-filled depressions. Male emits a short, *yap*-like call. Female lays around 1,100 eggs in a clutch, with development completed in about 7 weeks.

Aaron Payne

Blacksoil Toadlet ■ *Uperoleia trachyderma* TL 25mm

DESCRIPTION Body dorsolaterally compressed. Orange-brown to dark grey above, sometimes heavily mottled, and often with dull reddish vertebral stripe. Underparts white with grey flecking. Throat darker in breeding males. Hindlimbs short and toes without webbing, and with bright orange-red patches on groin and rear of thighs. **DISTRIBUTION** Found from Daly Waters region, NT, to McKinlay, Qld. **HABITS AND HABITAT** Nocturnal. Lives in grassy floodplains and claypans, usually on cracking black soil, emerging after heavy rains to feed and breed. Male's call is a short *cric*, repeated every 0.7 of a second. Reproductive biology poorly known.

Aaron Payne

Tyler's Toadlet ▪ *Uperoleia tyleri* TL 36mm

DESCRIPTION Generally grey to brown above with darker patches, and typically with pale triangular crown between eyes and nostrils. Covered with numerous short tubercles, occasionally tipped with orange, yellow or brown, and has large, prominent parotoid glands. Upper arms bright yellow, legs have darker barring, and groin and rear of thighs have yellow to orange-red patches. Underparts bluish-grey with black marbling and white spots. Very similar to Martin's Toadlet (see p. 97), but differs by call, which is shorter, and colouring of underside (blackish in Martin's Toadlet). **DISTRIBUTION** Found from near Genoa, Vic, to Sydney, NSW. Older texts suggest the species occurs in southern Vic west to Marlo; subsequent genetic analysis suggest these specimens including those from type series are attributable to Martin's Toadlet. **HABITS AND HABITAT** Nocturnal. Inhabits swamps and dams in coastal dunes, heaths and closed woodland. Male's call consists of a single *wra*, repeated about 3 seconds apart, given from beneath grass tussocks and sedges near the water's edge. Tadpoles take about 14 weeks to complete metamorphosis.

Grant Webster

PELODRYADIDAE (AUSTRALIAN TREE FROGS)
The tree frogs are variable in ecology and reproduction, from burrowing species to arboreal stream dwellers. The family (formerly the Hylidae) includes some of the world most recognizable species, such as the Green Tree Frog. Australian tree frogs are now in the Pelodryadidae family, following a revision of the world's tree frogs. This has gained acceptance among many herpetologists, although the taxonomy below this is somewhat unresolved. The genus *Litoria* is used below for consistency, however, is currently paraphyletic. Work currently being undertaken on the group will render significant changes to the taxonomy (see p. 169 for table summarizing the changes).

Striped Burrowing Frog ■ *Cyclorana alboguttata* TL 82mm

DESCRIPTION Colouration varies widely from brown-grey to greenish above, with or without mottling. Prominent yellow to green stripe extends from snout down spine to about the hips, and thighs have yellow to white spots on rear. Back mainly smooth. Underside whitish. Juveniles appear similar to adults. **DISTRIBUTION** Found from northern NSW (west of Great Dividing Range), to northeastern SA, and over much of Qld. **HABITS AND HABITAT** Lives around watercourses, including rivers and creeks,

pools and small waterbodies, where it feeds on invertebrates. Often seen basking in vegetation. Male's call, a single quack-like note repeated every few seconds, is made from edge of the water during spring and summer, especially after rain. Breeds in pools beside waterbodies or in still sections of a waterbody itself, and lays a clutch of about 2,500 eggs in a series of clumps. Tadpoles turn into frogs in about 14 weeks.

Giant Frog ■ *Cyclorana australis* TL 110mm

DESCRIPTION Colouration varies widely, from whitish-grey to orange-brown or green above, with black markings, to chocolate-brown, with or without mottling. Underparts whitish. Back mainly smooth and juveniles appear similar to adults. **DISTRIBUTION** Occurs from northwestern Qld, west to about Port Hedland, WA. **HABITS AND**

HABITAT Lives around watercourses, including rivers and creeks, pools and small waterbodies, where it feeds on invertebrates and other frogs. Male's call is a single note, sounding like a deep *honk*, repeated every few seconds, which is given from the water's edge during spring and summer, often after rain. Breeds in pools beside waterbodies or in still sections of a waterbody itself, and lays a clutch of about 6,000 eggs in a series of clumps. Tadpoles turn into frogs in about 6 weeks.

Short-footed Frog ■ *Cyclorana brevipes* TL 46mm

DESCRIPTION Colouration brown to grey or yellow with either lighter or dark patches. Usually a vertebral stripe. Back mainly smooth. Underside whitish. Juveniles appear similar to adults. **DISTRIBUTION** Found from northern NSW, to gulf region of Qld. **HABITS AND HABITAT** Lives around watercourses, including rivers and creeks, as well as pools

and small waterbodies. Feeds on invertebrates and other frogs. Male's call is made from the water's edge during spring and summer, frequently after rain. Call is a single note sounding like *waaaarr*, repeated every few seconds. Breeds in pools beside waterbodies or in still sections of a waterbody itself. Lays a clutch of about 2,000 eggs in a series of clumps. Tadpoles turn into frogs in about 4 weeks.

Hidden-ear Frog ■ *Cyclorana cryptotis* TL 47mm

DESCRIPTION Colouration pale grey to brown with darker mottling. Usually 5 orange stripes that can be prominent on some individuals, and sometimes a thin vertebral stripe. Back mainly smooth. Underside whitish. Juveniles appear similar to adults. Tympanum covered by skin. **DISTRIBUTION** Found in northern Australia, from Cape York to Dampier, WA. **HABITS**

AND HABITAT Lives around watercourses, including rivers and creeks, as well as pools and small waterbodies. Feeds on invertebrates and other frogs. Male's call is made from the water's edge during spring and summer, frequently after rain. Call is a single note sounding like *waaarr*, repeated every few seconds. Breeds in pools beside waterbodies or in still sections of a waterbody itself. Lays a clutch of about 900 eggs in a series of clumps. Tadpoles turn into frogs in about 4 weeks.

Angus McNab

Knife-footed Frog ■ *Cyclorana cultripes* TL 55mm

DESCRIPTION Pale grey to brown above, with darker mottling, and some individuals have a greenish tinge. Usually a narrow pale vertebral stripe. Back mainly smooth, with small, raised warts, which in some individuals form a weak dorsolateral fold. Underside whitish. Juveniles appear similar to adults. **DISTRIBUTION** Found inland in eastern and central Australia west of Great Dividing Range, from Dalby, Qld, west to Kununurra, WA.

HABITS AND HABITAT Lives around watercourses, including rivers and creeks, as well as pools and small waterbodies, where it feeds on invertebrates and other frogs. Male's call, a single note sounding like *waaannp*, repeated every few seconds, is made from the water's edge during spring and summer, frequently after rain. Breeds in pools beside waterbodies or in still sections of a waterbody itself. Tadpoles turn into frogs in about 4 weeks.

Scott Eipper

Long-footed Frog ■ *Cyclorana longipes* TL 56mm

DESCRIPTION Brown or grey above with darker mottling; some individuals paler with minimal dark markings. Back mainly smooth, with small, raised warts. Underside whitish. Juveniles appear similar to adults. **DISTRIBUTION** Found in northern Australia, from Mt Isa, Qld, to Dampier, WA. **HABITS AND HABITAT** Lives around watercourses, including

rivers and creeks, as well as pools and small waterbodies, where it feeds on invertebrates and other frogs. Male's call is a single note sounding like *waaarrrnnn*, repeated every few seconds, and is given from the water's edge the during spring and summer, most frequently after rain. Breeds in pools beside waterbodies or in still sections of a waterbody itself. Lays around 1,500 eggs in small clumps. Tadpoles turn into frogs about 4 weeks after hatching.

Daly Waters Frog ■ *Cyclorana maculosa* TL 55mm

DESCRIPTION Brown to grey or yellow above, with either lighter or dark mottling, and typically with vertebral stripe. Back mainly smooth. Underside whitish. Juveniles appear similar to adults. **DISTRIBUTION** Found in gulf region of Qld, across NT, into Kimberley region, WA. **HABITS AND HABITAT** Lives around watercourses, including

black soil floodplains and creeks, as well as pools and small waterbodies. Feeds on invertebrates and other frogs. Male's call is made from the water's edge during spring and summer, frequently after rain. Call is a single note sounding like *waaaarree*, repeated every few seconds. Breeds in pools beside waterbodies or in still sections of a waterbody itself. Lays eggs in a series of clumps. Tadpoles turn into frogs in about 2 weeks.

Main's Frog ■ *Cyclorana maini* TL 48mm

DESCRIPTION Brown or grey above, with darker mottling and dark stripe from nostril to flanks. Back mainly smooth, with small, raised warts. Underside whitish. Juveniles appear similar to adults. **DISTRIBUTION** Found in western Qld, across central Australia to about Exmouth, WA. **HABITS AND HABITAT** Lives around watercourses, including rivers and creeks, as well as pools and small waterbodies. Feeds on invertebrates and other frogs. Male's call is made from the water's edge during spring and summer, frequently after rain. Call is a single note sounding like *waaarr*, repeated every few seconds. It has been likened to that of a bleating sheep. Breeds in pools beside waterbodies or in still sections of a waterbody itself. Lays eggs in a series of clumps. Tadpoles turn into frogs in about 2 weeks.

Scott Eipper

Small Frog ■ *Cyclorana manya* TL 33mm

DESCRIPTION Brown to grey or yellow, with either lighter or dark mottling, and usually with vertebral stripe. Typically has pale broad collar running across body at shoulders. Back mainly smooth. Underside whitish. Juveniles appear similar to adults. **DISTRIBUTION** Found in Gulf region of Qld to WA. **HABITS AND HABITAT** Lives around watercourses, including black soil floodplains and creeks, as well as pools and small waterbodies, where it feeds on invertebrates and other frogs. Male's call is made from the water's edge during spring and summer, frequently after rain. Call is a single note sounding like *waaaa*, repeated every few seconds, not dissimilar to the cry of a baby. Breeds in pools beside waterbodies or in still sections of a waterbody itself. Lays eggs in a series of clumps. Tadpoles turn into frogs in about 3 weeks.

Grant Webster

New Holland Frog ■ *Cyclorana novaehollandiae* TL 105mm

DESCRIPTION Colouration varies widely from whitish-grey, to green, to chocolate-brown, with or without mottling. Back mainly smooth. Underside whitish. Juveniles appear similar to adults but are usually green. **DISTRIBUTION** Found from northern NSW (west of Great Dividing Range), to northeastern SA and over much of Qld. **HABITS AND HABITAT**

Lives around watercourses, including rivers and creeks, as well as pools and small waterbodies, where it feeds on invertebrates and other frogs. Male's call is made from the water's edge during spring and summer, frequently after rain. Call is a single note sounding like a deep *honk*, repeated every few seconds. Breeds in pools beside waterbodies or in still sections of a waterbody itself. Lays a clutch of around 4,900 eggs in a series of clumps. Tadpoles turn into frogs in about 5 weeks.

Western Water-holding Frog ■ *Cyclorana occidentalis* TL 66mm

DESCRIPTION Colouration orange-brown with or without mottling. Back mainly smooth. Underside whitish. Juveniles appear similar to adults. **DISTRIBUTION** Found over much of inland WA, excluding south and north of the state. **HABITS AND HABITAT** Lives around watercourses, including rivers and creeks, as well as pools and small waterbodies. Feeds on invertebrates and other frogs. Male's call is made from the water's edge during spring and

summer, frequently after rain. Call is a single note sounding like *waaaarrp*, repeated every few seconds. Thought to breed in pools beside waterbodies or in still sections of a waterbody itself. Breeding biology of this recently described species is thought to be similar to that of the Eastern Water-holding Frog (see opposite).

Eastern Water-holding Frog ■ *Cyclorana platycephala* TL 72mm

DESCRIPTION Plump frog with relatively small eyes. Usually green, grey or brown above, flecked darker, and pale cream below. Often has some scattered raised tubercles on back, but skin otherwise smooth, and toes webbed. **DISTRIBUTION** Found in inland Qld, NSW and across Barkly Tableland, NT. **HABITS AND HABITAT** Occurs in swamps, floodplains and claypans that are filled irregularly by rains. Possibly a species complex. During dry periods, lives in deep burrows in a cocoon it forms to protect itself from water loss, emerging only when sufficient surface water is present to enable it to breed and lay its eggs in the temporary pools that have formed. Large numbers of eggs are produced at this time, often 1,700 or more, and the tadpoles must hatch and develop quickly, before the pools dry up. Tadpoles turn into frogs in about 5 weeks.

Northern population

Wailing Frog ▪ *Cyclorana vagita* TL 49mm

DESCRIPTION Colouration pale grey to brown with darker mottling. Some individuals have a greenish tinge. Usually a narrow pale vertebral stripe. Back mainly smooth, with small, raised warts, which in some individuals form a weak dorsolateral fold. Underside whitish. Juveniles appear similar to adults. **DISTRIBUTION** Found in Kimberley region of

WA and neighbouring NT. **HABITS AND HABITAT** Lives around flooded grassland, claypans and small bodies of still water. Feeds on invertebrates and other frogs. Call is a single note sounding like *waaarrrrrr*, repeated every few seconds. Breeds in pools beside waterbodies or in still sections of a waterbody itself. Tadpoles turn into frogs in about 3–4 weeks.

Rough Frog ▪ *Cyclorana verrucosa* TL 49mm

DESCRIPTION Colouration pale yellow to brown with darker green mottling. Usually a narrow pale vertebral stripe. Back mainly smooth, with small, raised warts. Underside whitish. Juveniles appear similar to adults. **DISTRIBUTION** Found inland in southern Qld and northern NSW west of Great Dividing Range. **HABITS AND HABITAT** Lives around watercourses, including pools and small waterbodies, as well as flooded grassland

and claypans. Feeds on invertebrates and other frogs. Male's call is made from the water's edge during spring and summer after rain. Call is a single note sounding like *waaarnp*, repeated every few seconds. Breeds in pools beside waterbodies or in still sections of a waterbody itself. Tadpoles turn into frogs in about 5 weeks.

Slender Tree Frog ■ *Litoria adelaidensis* TL 49mm

DESCRIPTION Colouration varies from light to dark brown or sometimes green, with dark brown stripe edged with white on lower edge. Some individuals flecked with black. Back smooth, without bumps or protrusions. Backs of thighs have red spots or can be entirely red. **DISTRIBUTION** Found in southern WA. **HABITS AND HABITAT** Occupies a wide range of habitats, including artificial and natural watercourses such as

dams, swamps, ephemeral pools, areas beside creeks and streams, and roadside ditches. Feeds on invertebrates. Male's call is made while sitting on emergent plants, including reeds, or on water lilies. Call is a drawn-out *screeeeech*. Spawn is a series of small clusters, each consisting of about a dozen eggs, which are attached to the emergent stem of a water plant. Can produce up to 1,363 eggs in a clutch, becoming frogs about 12–15 weeks later.

Scott Eipper

Cape Melville Tree Frog ■ *Litoria andiirrmalin* TL 110mm

DESCRIPTION Colouration variable, but usually brown to grey with darker mottling, and most individuals have white to gold blotching randomly scattered on dorsum. White to cream below, with brown throat on breeding males. **DISTRIBUTION** Only found around Cape Melville, northern Qld.

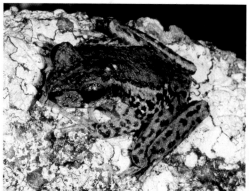

HABITS AND HABITAT Occurs along streams and creeks in rainforests and vine forests. Known to inhabit areas of moving water, such as cascades and riffles among large granite boulders. This large frog was described in 1997 and is fairly poorly known. It is thought to eat invertebrates. Its call sounds like a rapidly repeated *pop-pop-pop-pop-pop*. Reproductive biology poorly known.

H. B. Hines/DES

Green and Golden Bell Frog ■ *Litoria aurea* TL 106mm
(Golden Bell Frog; Green and Gold Swamp Frog)

DESCRIPTION Colouration variable, but typically gold with large green blotches and spots, although some individuals are almost all gold or all green. Colouration can change depending on how warm the frog is, being brightest when it is at its preferred body temperature. Back essentially smooth, without warty bumps or protrusions. **DISTRIBUTION** Found in southeastern Australia, from Orbost, Vic, to near Byron Bay, northern NSW. **HABITS AND HABITAT** Favours swamps, large dams and lagoons, and also uses southern wallum swamps. Has suffered major decline due to chytrid fungus and

habitat loss. Males often call from floating vegetation and among reeds near the water level from spring to autumn, primarily after rain. Call has been described as a single, drawn-out *waaark*. Spawn may be seen in floating clusters, but can also be submerged and wrapped around aquatic plants. Average clutch numbers 5,120 eggs. Tadpoles take about 12 weeks to complete metamorphosis.

Kimberley Rockhole Frog ■ *Litoria aurifera* TL 22mm

DESCRIPTION Mottled grey or brown above, depending on rock colour it is found on. Limbs banded in irregular manner. Small, raised tubercles across body help to disrupt the frog's shape, aiding in camouflage. Underside white to cream. Very similar to the Rockhole Frog (see p. 136), differing by having a pointed snout and a different call. **DISTRIBUTION** Restricted to Kimberley region, WA. **HABITS AND HABITAT**

Restricted to rocky gorges that have both permanent and season-dependent watercourses. This tiny species is different from most Australian frogs in that it is diurnal. Its small, agile body allows it to escape predators. It was described in 2010 and is quite poorly known. It is thought to eat invertebrates, and its call sounds like a rapidly repeated *pip-pip-pip-pip*. Produces about 77 eggs in a clutch.

Kimberley Rocket Frog ■ *Litoria axillaris* TL 26mm

DESCRIPTION Grey or brown above with broad black stripe extending from nose through eye and above ear. It continues lower, from armpit along flank. Body smooth without tubercles. Underside white to cream. Very similar to Tornier's Frog (see p. 154), differing by having a dorsolateral stripe.

DISTRIBUTION Restricted to western Kimberley region, WA. **HABITS AND HABITAT** Restricted to rocky gorges that have both permanent and season-dependent watercourses. This small frog was described in 2011 and is quite poorly known. It is thought to eat invertebrates. Its call sounds like a rapidly repeated *pop-pop-pop-pop-pop*. Reproductive biology poorly known.

Brad Maryan

Slender Bleating Tree Frog ■ *Litoria balatus* TL 42mm

DESCRIPTION Cream to dark brown, with darker irregular stripe running from nose over shoulders down to near hips. Remaining areas usually light brown to grey, lightening on to lower flanks. Underside cream to white. Back smooth, without bumps or protrusions. Differs from both the Bleating Tree Frog (see p. 127) and Screaming Tree Frog (see p. 145) by dorsolateral line that continues above arm and more slender body form. **DISTRIBUTION** Found along coast and associated ranges, QLD, from Bundaberg to NSW border region. Precise southerly distribution limits currently unclear. **HABITS AND HABITAT** Occurs in

moist habitats such as swamps, rainforests, eucalypt forests, heaths and wallum. Also utilizes urban environments, including toilets, roadside ditches, cattle troughs and dams. Feeds on invertebrates. Male's call made from vegetation and at bases of grass clumps around a waterbody. Call long and sounds like a bleating sheep. Lays an average clutch of 1,100 eggs, which once hatched take about 8 weeks to turn into juvenile frogs.

Scott Eipper

Barrington Tops Tree Frog ■ *Litoria barringtonensis* TL 29mm

DESCRIPTION Bright green above, with dark-edged gold or yellow stripe extending from snout through eye and on to flanks. Small black flecks sparsely distributed across back. Underside white to cream. Back smooth, without bumps or protrusions. DISTRIBUTION Found in eastern NSW, from Barrington Tops north to Gibraltar Range. HABITS AND

HABITAT Lives around creeks and streams in forests, breeding in ephemeral pools beside creeks and streams and in slower parts of the main waterway. Feeds on invertebrates. Male's call is made from overhanging vegetation or branches usually about 30cm above the water level. Call is made in two parts – the first a *wrekik-wrekik* and the second a loud *eeep*, which is not always repeated. Average clutch size is about 450 eggs. Tadpoles transform to frogs in about 8–10 weeks.

Beautiful Tree Frog ■ *Litoria bella* TL 42mm

DESCRIPTION Colouration varies from light to dark green. Back slightly rough, made up of granular, tiny warty bumps or protrusions. Back of thigh usually blue, maroon or orange. Similar to the Dainty Green Tree Frog (see p. 130), but lacks yellow stripe from

nostril to ear. Underside yellow to orange. DISTRIBUTION Found on Cape York Peninsula, Qld. HABITS AND HABITAT Seeks out small waterbodies in rainforests and vine forests, such as small billabongs, as well as ephemeral pools beside rivers and creeks. Feeds on invertebrates. Male's call is made from trees and shrubs surrounding the water, particularly after rain, in spring to early autumn. Call is a drawn-out *wahh* that lasts about a second or so and is repeated for extended periods. One of Australia's newest frog species, it has been known to frog researchers for some time but has only recently been described. Average clutch size is 844 eggs. Tadpoles transform into frogs in about 8 weeks.

Northern Dwarf Tree Frog ■ *Litoria bicolor* TL 26mm

DESCRIPTION Colouration varies from light to dark green or brown, with white to cream flanks. Dark stripe extends from snout, along side to just above front legs. Some individuals flecked with black. Underside cream to white. Back smooth, without bumps or protrusions. **DISTRIBUTION** Found in northern Australia, from northern Qld across to about Broome, WA. **HABITS AND HABITAT** Occupies wide range of habitats, including artificial

and natural watercourses such as dams, swamps, ephemeral pools, areas beside creeks and streams, and roadside ditches. Feeds on invertebrates. Male's call – a *weee-kip* in two parts – is made while sitting on emergent plants, including reeds, or on water lilies. Spawn is a series of small clusters, each consisting of about 15 eggs, which are attached to the emergent stem of a water plant. Produces up to 1,000 eggs in a clutch, becoming frogs about 11 weeks later.

Scott Eipper

Booroolong Frog ■ *Litoria booroolongensis* TL 47mm

DESCRIPTION Colouration varies from light tan to dark grey, with or without flecks of black and darker brown. Thin dark stripe extends along canthal ridge across top of tympanum. Underside cream to white. Back has bumps or protrusions. **DISTRIBUTION** Found in southeastern Australia, from Qld border along Great Dividing Range down to just over the Vic border. **HABITS AND HABITAT** Occupies natural watercourses such as rocky streams, river headwaters and creeks. Often seen on rocks around edges of water. Has

suffered major decline due to chytrid fungus, habitat loss and trout introduction. Feeds on invertebrates. Male has a soft call that sounds like *wrk-wrk-wrk-wrk*, and is made while sitting on emergent rocks and stones in the water. Stream breeder, laying about 1,300 eggs that are adhered to rocks in a shallow section of a stream. When these hatch, tadpoles take about 11 weeks to become frogs.

Adam Elliott

Green-thighed Frog ■ *Litoria brevipalmata* TL 60mm

DESCRIPTION Colouration varies from reddish-brown to dark grey, with or without flecks of black and darker brown. Usually a conspicuous white lip. Thick dark stripe extends from nostril through eye down on to flanks. On lower part of flanks it becomes spots and blotches encased in a green-yellow band. Breeding males become yellowish. Underside cream to white. Backs of thighs bright green. **DISTRIBUTION** Found in southeastern Qld, along coast down to Ourimbah, NSW. **HABITS AND HABITAT** Occupies lowland forests, heath and wallum with watercourses such as swamps, ponds and dams. Often seen after

very heavy downpours, possibly because it spends much of its time buried beneath the ground up to 1m deep. Has suffered decline due to habitat loss. Feeds on invertebrates. Male has a short call that sounds like *quack–quack-quack-quack* and speeds up towards the end of each call burst. Pond breeder. Spawn of about 400 eggs is adhered to vegetation. When eggs hatch, tadpoles take about 10 weeks to become frogs.

Tasmanian Tree Frog ■ *Litoria burrowsae* TL 62mm

DESCRIPTION Colouration varies from green to dark grey, with flecks of black, white and brown. This can be quite extensive, giving the appearance of marbling in some individuals. Underside pink to white. Backs of thighs dull brown. **DISTRIBUTION** Found in western

Tas. **HABITS AND HABITAT** Occupies rainforests, heath, alpine grassland, swamps, ponds and dams. May be suffering declines due to chytrid fungus and habitat loss. Feeds on invertebrates. Males call in a series of short *hoonks* after heavy rain, from trees and vegetation around waterbodies. Pond breeders. Spawn of about 100 eggs is adhered to submerged vegetation. When eggs hatch, tadpoles take about 30 weeks to become frogs.

Green Tree Frog ■ *Litoria caerulea* TL 102mm
(White's Tree Frog, Dumpy Tree Frog)

DESCRIPTION Colouration varies, although typically light to dark green, but can be brown or completely blue. Some individuals are a solid colour, while others have varying amounts of white flecks and blotches. Back smooth, without bumps or protrusions. **DISTRIBUTION** Found from NSW to WA. **HABITS AND HABITAT** Occurs in grassland, forests to deserts, and can thrive in urban environments. Seeks out waterbodies, both natural and artificial, including roadside ditches and dams, as well as ephemeral pools beside rivers, streams and creeks. Feeds on invertebrates and smaller frogs, but has also been recorded eating small snakes, geckos and bats. Very popular as a pet. Males call

mainly between spring and early autumn from inside hollow logs, downpipes and the ground near the edge of a waterbody, particularly just before or after rain. Call is a deep, resonating *brawwk*. Spawn is a series of large clumps that float on the surface, but occasionally it is attached to emergent vegetation. Number of eggs can vary greatly – usually about 2,500 eggs in a clutch. Tadpoles can take 6 weeks to complete metamorphosis.

Scott Eipper

Scott Eipper

Yellow-spotted Tree Frog ■ *Litoria castanea* TL 98mm

DESCRIPTION Colouration varies, but usually gold with large green blotches and spots, and in some individuals almost all gold or all green. Depending on a frog's warmth, colouration can change, being brightest when it is at its preferred body temperature. Back has wart-like bumps or protrusions. Groin and backs of thighs adorned with conspicuous yellow spots. **DISTRIBUTION** Was found in southeastern Australia, from New England tablelands south to near Yass, NSW. Individuals fitting appearance of this species occur in New Zealand. **HABITS AND HABITAT** Possibly synonymous with northern form of the Southern Bell Frog (see p. 146). Has suffered massive decline due to chytrid fungus, habitat

loss and feral fish introduction. It may have completely disappeared in the wild. Feeds on invertebrates. Prefers swamps, large dams or lagoons, and other slow-moving or still waterbodies. Male's call made from among floating vegetation and on reeds near the water level. Call has been described as a single, drawn-out *crawark-crawark-crok-crok*. Reproductive biology poorly known.

Cave-dwelling Frog ■ *Litoria cavernicola* TL 67mm

DESCRIPTION A robust frog. Colouration varies from light to dark green or olive. Back granular without bumps or protrusions. Juveniles similar to adults but can have white flecking. **DISTRIBUTION** Found in northern WA. **HABITS AND HABITAT** Shelters in caves and

rock escarpments. Lives in cooler gorges, emerging at night to feed on walls and surrounding nearby pools. Feeds on invertebrates. Male's call has been recorded at 1.5–2m above the ground in trees and on rock faces, primarily after rain, during summer. Call is a short *waark*. Spawn is laid in shallow rock pools. Tadpoles take an average of just 3 weeks to complete metamorphosis.

Red-eyed Tree Frog ■ *Litoria chloris* TL 65mm

DESCRIPTION Colouration varies from light to dark green, occasionally with yellow spots on back. Iris bright red to orange. Backs of thighs bright blue, which is the main difference between this species and the Orange-thighed Frog (see p. 157). Back granular. **DISTRIBUTION** Found across northeastern Australia, from Nowra, NSW, north to Proserpine, Qld. **HABITS AND HABITAT** Seeks out small waterbodies in forests and gardens, where it feeds on invertebrates. Males often call from vegetation including branches in trees and shrubs, and the ground surrounding a waterbody. Call, made particularly just before or after rain, is made in two parts – the first consists of a drawn-out *wahh wahh wahh* and the second is a soft trill, which is not always repeated. Spawn is a single sheet or broad clusters of eggs laid near the water's edge, often attached to emergent vegetation. A single spawning can comprise a clutch of about 1,350 eggs. Tadpoles can take 6 weeks to complete metamorphosis. Some metamorphlings are yellow to light brown in colour, while others are green.

Scott Eipper

Blue Mountains Tree Frog ■ *Litoria citropa* TL 65mm

DESCRIPTION Colouration varies but is usually brown on top with green and white to cream on flanks. Pink to red between shoulders to base of hips. Dark stripe edged with white-cream extends from snout, along side to just above front legs; green below this. Some individuals flecked with black or dark brown. A reduced brown phase that is almost metallic bluish-green is displayed in captivity. Back smooth, without bumps or protrusions. **DISTRIBUTION** Found in southeastern Australia, from Wollemi National Park, NSW, to Gippsland, northeastern Vic. **HABITS AND HABITAT** Lives around creeks and streams in forests. Suffering from decline due to chytrid fungus and habitat loss. Feeds on invertebrates. Male's call is made from the water's edge, on rocks in the water or from overhanging vegetation. Call is in two parts – the first a *waaark* and the second a gurgling bobble repeated quickly. Can lay an average clutch size of 1,370 eggs. Tadpoles transform to metamorphlings 9–18 weeks later, depending on water and air temperatures.

Scott Eipper

Cooloola Sedge Frog ■ *Litoria cooloolensis* TL 31mm

DESCRIPTION Colouration varies from light to dark green or tan. White stripe extends along top lip to front edge of shoulder. Usually fine black dots covering body. Some individuals flecked with black. Back smooth to finely granular. Underside whitish to yellow. **DISTRIBUTION** Found on south-east Qld coast, from North Stradbroke Island to Fraser Island. **HABITS AND**

HABITAT Occupies wallum swamps and surrounding forests. Feeds on invertebrates. One of four of the acid frogs, this species is endangered. Male's call is made while it sits on emergent plants, including reeds, or on sedges. Call sounds like *eeeeee-kik*, in two parts. Reproductive biology thought to be similar to that of the Wallum Sedge Frog (see p. 141).

Copland's Rock Frog ■ *Litoria coplandi* TL 41mm

DESCRIPTION Colouration mottled grey or brown depending on colour of rock it is found on. Flanks usually lighter brown or pinkish. Body finely granular and depressed. Underside white to cream. Usually prominent dark webbing between toes. **DISTRIBUTION** Found from around Winton, Qld, across to Kimberley region, WA. **HABITS AND HABITAT** Restricted to rocky gorges and environments that have both

permanent and season-dependent watercourses. Depressed body is a probable adaption to allow for shelter in rock crevices. Thought to eat invertebrates. Call made while sitting on rocks and branches in and surrounding waterbody. Call sounds like rapidly repeated chirps. Produces 350 eggs in a clutch laid on small rocks in pools. Tadpoles take about 8 weeks to become frogs.

Spotted-thighed Frog ■ *Litoria cyclorhyncha* TL 70mm

DESCRIPTION Colouration varies, but usually grey with large green blotches and spots. Some individuals are almost all grey or all green. Back essentially smooth, with warty bumps. Granular underside is whitish. Insides of thighs and arms jet black with conspicuous yellow, white or blue spots. **DISTRIBUTION** Found in coastal south WA, from Cheyne Beach to Israelite Bay. **HABITS AND HABITAT** Prefers swamps, large dams or lagoons. Males often call from floating vegetation and among reeds near the water level from spring to autumn, primarily after rain. Call has been described as a drawn-out *wahhh–wahhh-waaark*. Pond or pool breeder, preferring little or no surface movement. Spawn found in floating clusters, but may also be submerged and wrapped around aquatic plants. Tadpoles take about 10 weeks to complete metamorphosis.

Angus McNab

Dahl's Aquatic Frog ■ *Litoria dahlii* TL 72mm

DESCRIPTION Colouration varies, but usually green with darker blotches and spots, which might be brown, tan or greyish. Usually a yellow to green vertebral stripe. Some individuals are almost all gold or all green. Colouration can change depending on activity level of the frog. Back essentially smooth, without warty bumps or protrusions. Slimy to touch. **DISTRIBUTION** Found in northern Australia, from Kimberleys, WA, across to western Cape York Peninsula, Qld. **HABITS AND HABITAT** Prefers swamps, large dams, billabongs or lagoons, and will used flooded grassland. In the dry season it buries

itself between cracking soil to find shelter. Possibly more than one species. This one produces toxic mucus as a defensive strategy to ward off predators. Eats invertebrates and other frogs. Known to eat underwater – something not recorded in other Australian species. Males often call from floating vegetation after rain. Call is as a repeated, short *wrk-wrk-wrk*.

Davies' Tree Frog ■ *Litoria daviesae* TL 62mm

DESCRIPTION Colouration varies, but usually brown on top with green and white to cream on flanks. Lower flanks orange to pink. Lower rear flanks mottled black and white. Dark stripe edged with white-cream extends from snout, along sides to just above front legs; green below this. Some individuals flecked with black or dark brown. Back smooth, without bumps or protrusions. **DISTRIBUTION** Found in northeastern NSW in Hasting River catchment. **HABITS AND HABITAT** Lives around creeks and streams in forests, breeding in ephemeral pools beside creeks and streams, and in slower part of main waterway. Feeds on invertebrates. Male's call is made from the water's edge, on rocks in the water or from overhanging vegetation, particularly after heavy rain, in spring to early summer. Call sounds like a *waahh rarr*, repeated every second or so. Lays an average clutch size of 400 eggs. Tadpoles transform to metamorphlings 9–18 weeks later, depending on water and air temperatures.

Grant Webster

Australian Lace Lid ■ *Litoria dayi* TL 53mm

DESCRIPTION Colouration varies from light brown to dark brown, occasionally yellowish. Can have black variegations as well white spotting, but this does not occur in all individuals. Large black eyes, and superficially looks very similar to the lace-lids of PNG. Lower eyelid translucent with conspicuous yellow markings. Back slightly rough and granular. Underside white to yellow. **DISTRIBUTION** Found in wet tropics region of northeastern Qld. **HABITS AND HABITAT** Seeks out streams and ephemeral pools in

rainforests. Often seen sitting on rocks in the middle of a waterway at night. Has suffered significant decline due to chytrid fungus and habitat loss. Feeds on invertebrates. Male's call is made from trees and shrubs surrounding the water. Call is a *werrerec* that lasts about a second and is repeated for extended periods. Average clutch size of 210 eggs is laid in clumps. Tadpoles transform to become frogs about 16 weeks later.

Bleating Tree Frog ■ *Litoria dentata* TL 42mm

DESCRIPTION Colouration varies from cream to dark brown, with darker irregular stripe running from nose, over shoulders, down to near hips. Remaining areas usually light brown to grey, lightening on to lower flanks. Underside cream to white. Back smooth, without bumps or protrusions. Differs from Slender Bleating Tree Frog (see p. 115) by having a dorsolateral line posterior to forelimb and more robust body form. Differs from Screaming Tree Frog (see p. 145) by having dark vocal sac vs. yellow and a more northerly distribution. **DISTRIBUTION** Found along GDR from Taree, NSW, to Qld/NSW border. May occur in Qld but further genetic sequencing of Main Range/Gold Cost specimens not complete. **HABITS AND HABITAT**

Occurs in moist habitats such as swamps, rainforests, eucalypt forests, heaths and wallum. Also utilizes urban environments, including toilets, roadside ditches, cattle troughs and dams. Feeds on invertebrates. Male's call is made from vegetation and at bases of grass clumps around a waterbody. Call is long and sounds like a bleating sheep. Reproductive biology likely similar to Screaming Tree Frog.

Buzzing Tree Frog ■ *Litoria electrica* TL 39mm

DESCRIPTION Colouration varies from cream to dark brown, with dark brown to black stripe that runs along length of body; bottom edge of stripe often merges with side colouration on to white ventral surface. Back smooth, without bumps or protrusions. Usually a pair of darker wide, irregular bands across hips. **DISTRIBUTION** Found in western Qld into base of Gulf of Carpentaria. **HABITS AND HABITAT** In dry habitats such as deserts, Mitchell grass downs and savannah, seeks out small waterbodies, both

natural and artificial, including toilets, roadside ditches, cattle troughs and dams, as well as ephemeral pools beside rivers and creeks. Congregates around drying pools in seasonal waterways. Feeds on invertebrates. Male's call is made from the ground around a waterbody or in low vegetation, primarily after rain. Call sounds very similar to buzzing noise given off by an electric arc. Reproductive biology probably similar to that of the Desert Tree Frog (see p. 150).

Growling Tree Frog ■ *Litoria eucnemis* TL 72mm

DESCRIPTION Colouration varies from reddish-brown to grey-brown. Heavily mottled with greens, browns, greys and yellows, with black flecking, resembling mosses and lichens of a rainforest tree trunk. Legs have broad bands. Back slightly rough and granular. Edges of legs have tubercles to assist in breaking up the body shape (to confuse potential predators). Underside whitish. Upper iris turquoise. **DISTRIBUTION** Occurs in Wenlock River area

of northeastern Qld and southern PNG. **HABITS AND HABITAT** Usually found in rainforests and vine thickets along waterways and seepages. By day sits against trees, relying on its excellent camouflage. Feeds on invertebrates. Male's call is made from trees and shrubs surrounding water. Call has two parts – the first is a series of rapid clicks that can speed up to form a purring growl lasting about a second. Reproductive biology probably similar to that of the Green-eyed Tree Frog (see p. 151).

Brown Tree Frog ■ *Litoria ewingii* TL 41mm

DESCRIPTION Colouration varies from light grey to green, with dark brown to tan stripe that extends from snout through eye and along flanks. Some individuals can have additional stripes and dark flecks. Insides of hip and groin light orange without black markings. Back smooth, without bumps or protrusions. **DISTRIBUTION** Found in Vic, Tas, SA and southern NSW. **HABITS AND HABITAT** Lives around slow-moving or still watercourses, including artificial dams, lakes and ornamental pools. May be a species complex. Feeds on invertebrates. Call is a drawn-out *creeee creeee creeee*, repeated quite frequently. Clutch size can be about 250 eggs, spread across a dozen floating groups.

Dwarf Tree Frog ■ *Litoria fallax* TL 27mm
(Eastern Sedge Frog)

DESCRIPTION Colouration varies from light to dark green, with white to cream flanks, and dark stripe that runs from snout, along sides to just above front legs. Some individuals are flecked with black. Back smooth, without bumps or protrusions. **DISTRIBUTION** Found in eastern Australia, from northern Qld to southern NSW; also an introduced population in Melbourne, Vic. **HABITS AND HABITAT** Occupies wide range of habitats, including artificial and natural watercourses such as dams, swamps, ephemeral pools, areas

beside creeks and streams, and roadside ditches. Feeds on invertebrates. Male's call is made while sitting on emergent plants, including reeds, or on water lilies. Call is a *weee-kip* in two parts, and is intensified after rain. Pond or pool breeder, but also breeds in ephemeral pools alongside streams, preferring little or no surface movement. Spawn is a series of small clusters, each consisting of about 12 eggs, which are attached to the emergent stem of a water plant. Produces about 1,200 eggs in a clutch, becoming frogs about 10–15 weeks later.

Scott Eipper

Wallum Rocket Frog ■ *Litoria freycineti* TL 42mm

DESCRIPTION Mottled mix of browns, reds, yellows, greys and whites above, often with blotching and stripes, and with dark stripe extending from snout, along sides to just above front legs. Some individuals flecked with black. Back covered in bumps or protrusions that are not aligned in longitudinal rows. Underparts whitish. **DISTRIBUTION** Found in coastal regions of eastern Australia, from Fraser Island, Qld, to around Jervis Bay, NSW.

HABITS AND HABITAT Inhabits wallum swamps, heaths and neighbouring forests and grassland. A type species of the genus *Litoria*. Its coastal habitat preference is reflected in the meaning of *Litoria*, which is beach. Feeds on invertebrates. Male's call, a short cluck, sometimes preceded by a secondary whirring, is made while sitting around still waterbodies or on exposed points in a waterbody. Spawn is laid across base of a waterbody, producing about 450 eggs. Tadpoles develop into frogs in about 8 weeks.

Scott Eipper

Centralian Tree Frog ■ *Litoria gilleni* TL 81mm

DESCRIPTION Colouration variable. Typically light to dark green, but can be brownish, and some individuals are a solid colour, while others have varying amounts of white flecks and blotches. Back smooth, without bumps or protrusions. **DISTRIBUTION** Found from NSW to WA. **HABITS AND HABITAT** Occurs in gorges and localized forest associations,

where it inhabits waterbodies such as springs, rivers and seasonal creeks, feeding on invertebrates. Males call mainly between spring and early autumn from inside hollow logs and the ground near the edge of a waterbody, particularly just before or after rain. Call is a deep, resonating *brawwk*. Reproductive biology probably similar to that of the Magnificent Tree Frog (see p. 152).

Dainty Green Tree Frog ■ *Litoria gracilenta* TL 41mm

DESCRIPTION Colouration above varies from light to dark green, with yellow stripe that extends along head from nostrils, above eye and on to body. Back slightly rough, made up of granular, tiny warty bumps or protrusions. Backs of thighs usually blue or maroon. Underparts yellow to orange. **DISTRIBUTION** Found across northeastern Australia, from Sydney, NSW, extending north to Cooktown, Qld. Often occurs outside its natural range due to travelling in fruit shipments. **HABITS AND HABITAT** Uses small waterbodies,

both natural and artificial, including roadside ditches and dams, as well as ephemeral pools beside rivers and creeks. Feeds on invertebrates. Male's call, a drawn-out *wahh* that lasts about a second and is repeated for extended periods, is given from trees and shrubs surrounding the water, particularly after rain, in spring to early autumn. Tadpoles transform into frogs about 8 weeks after hatching.

Peters' Frog ■ *Litoria inermis* TL 42mm
(Bumpy Rocket Frog)

DESCRIPTION Colouration a mottled mix of browns, reds, yellows, greys and whites, often with faint blotching. Dark stripe extends from snout, along sides to just above front legs. Back covered in bumps or protrusions that are not aligned in longitudinal rows. Underside whitish. **DISTRIBUTION** Found in northern Australia, from Fraser Island across much of Qld to Kimberley region, WA. **HABITS AND HABITAT** Occurs in swamps, floodplains, savannah and open forests. Feeds on invertebrates. Male's call, a short *wreeeee* usually with short clucks at the end, is given while sitting around still waterbodies or on exposed points in a waterbody. Spawn is laid across base of a waterbody and contains about 300 eggs. Tadpoles develop into frogs in about 10 weeks.

Scott Eipper

Giant Tree Frog ■ *Litoria infrafrenata* TL 143mm
(White Lipped Tree Frog)

DESCRIPTION One of Australia's largest native amphibians. Colouration varies from light to dark green, as well as brown, with a white lip stripe that extends on to shoulder. Back smooth, without bumps or protrusions. **DISTRIBUTION** Occurs in coastal Qld, from Conway to tip of Cape York Peninsula, as well as PNG. **HABITS AND HABITAT** Seeks out waterbodies, both natural and artificial, including dams, ephemeral pools beside rivers, streams and creeks, where it feeds on invertebrates. Male's call is similar to that of a large dog's bark, repeated frequently, and is heard before or after rain, in spring to early autumn. Male calls from inside hollow logs, downpipes and exposed trees, before moving down to a waterbody. Spawn is a series of bell-shaped clumps that float on the water's surface, and a single spawning can comprise 4,000 eggs. Hatching occurs two days after laying at a water temperature of 24–30ºC, and it takes about 8 weeks for the tadpoles to become frogs.

Tyese Eipper

Jervis Bay Tree Frog ■ *Litoria jervisiensis* TL 64mm

DESCRIPTION Colouration varies from light grey to tan, with dark brown to tan stripe that extends from snout, through eye and along flanks. Some individuals can have additional stripes and dark flecks. Insides of armpits yellow, and groin pale orange. Back smooth, without bumps or protrusions. **DISTRIBUTION** Found from Vic border to Port Macquarie, NSW. **HABITS AND HABITAT** Lives around slow-moving or still

watercourses, including artificial dams, lakes and ornamental pools, where it feeds on invertebrates. Male's call, a pulsed *creeee creeee creeee*, repeated frequently, is made from grass tussocks, reeds, low, overhanging branches and emergent vegetation, from autumn to early spring. Eggs are laid in small floating groups, sometimes attached to plant material, with a clutch of about 900 eggs spread over a dozen or more floating groups. Tadpoles takes about 11 weeks to change into frogs.

Jungguy Tree Frog ■ *Litoria jungguy* TL 80mm

DESCRIPTION Dark brown above, with dark stripe that extends from snout, through eye and along flanks. Back smooth, without bumps or protrusions, and can be immaculate,

spotted or mottled. Rear of thighs spotted in light blue or yellow. Male exhibits distinct colour change to bright yellow when in breeding condition. This species cannot be split from Wilcox's Frog (see p. 157) by its morphology or call. **DISTRIBUTION** Occurs in northeastern Qld, from Tully to Cooktown, with an apparently isolated population at Eungella, further south. **HABITS AND HABITAT** Lives around creeks and streams in forests, breeding in ephemeral pools beside creeks and streams, and in slower parts of the main waterway, and feeding on invertebrates. Male's call, which is barely audible, is made from rocks in and on the edge of a waterway. Call is a soft, purring *bobble*, repeated quite frequently. Eggs are laid in a mass, loosely attached to the rocky base of a stream or in an ephemeral pool, in which it constructs a circular depression in the substrate to form a type of nest. Average clutch contains around 1,400 eggs, and hatched tadpoles start to develop into metamorphlings after about 12 weeks.

Kroombit Tree Frog ■ *Litoria kroombitensis* TL 35mm

DESCRIPTION Bright green to olive above, with dark-edged gold or yellow stripe extending from nostrils, above eyes, and broadening over flanks. Back smooth, without bumps or protrusions. Lip white, usually with broken section just before eye. Back often peppered with black flecks. Lower flanks often light grey-purple. Underside white.
DISTRIBUTION Only found in a few streams around Kroombit National Park, Qld.
HABITS AND HABITAT Lives around creeks and streams in forests, breeding in
ephemeral pools beside creeks and streams, and in slower parts of the main waterway. Feeds on invertebrates. Male's call is made from overhanging vegetation or branches usually about 30cm above the water level. Call is made in two parts – the first a loud *eeep*, followed by a deep *reeeeeck reeeck*. Often the sequence might be swapped or include slow clicks. Reproductive biology largely unknown, but thought to be similar to that of the Cascade Tree Frog (see p. 142).

Scott Eipper

Broad-palmed Frog ■ *Litoria latopalmata* TL 43mm

DESCRIPTION Colouration highly variable. Usually plain, pale brown to dark brown or grey. Other individuals mottled, with or without faint blotching. Dark stripe extends from snout, along sides to just above front legs – there is a conspicuous break before the eye in the stripe. Back smooth. Underside whitish. **DISTRIBUTION** Found over most of Qld and NSW; also eastern SA.
HABITS AND HABITAT
Occurs in swamps, floodplains, savannah and open forests. Feeds on invertebrates. Male's call is made while sitting around still waterbodies or on exposed points in a waterbody. Call is a short *cluck* that is repeated and gets faster as the call pesists. Spawn is laid across base of a waterbody, producing about 310 eggs.

Scott Eipper

Lesueur's Frog ■ *Litoria lesueuri* TL 74mm
(Rocky River Frog)

DESCRIPTION Dark brown above, with dark stripe that extends from snout, through eye and along flanks. Backs of thighs spotted in light blue. Male exhibits a distinct colour change to bright yellow when in breeding mode. Back smooth, without bumps or protrusions. **DISTRIBUTION** Found in southeastern Australia, from Blue Mountains, NSW, to west of Melbourne, Vic. **HABITS AND HABITAT** Lives around creeks and streams in forests, breeding in ephemeral pools beside creeks and streams, and in slower parts of

the main waterway. Feeds on invertebrates. Male's call, which is barely audible, is made from rocks in and on the edge of a waterway. Call is a soft, purring *bobble*, repeated quite frequently. Eggs are laid in a mass, loosely attached to the rocky base of a stream, ephemeral pool or creek, or to plant material. Average clutch size is 1,600 eggs. Tadpoles start to develop into metamorphlings after about 7 weeks.

Northern Heath Frog ■ *Litoria littlejohni* TL 72mm

DESCRIPTION Colouration varies from light grey to dark brown, with or without broad dark brown stripe that extends from behind eyes, over head and down back. Some individuals can have additional stripes and dark flecks. Insides of armpits and groin pale orange. Back smooth. **DISTRIBUTION** Found from Vic border to Watagan Ranges, NSW. Has suffered significant decline due to chytrid fungus and habitat loss; relatively low clutch size also results in slow population recovery. **HABITS AND HABITAT** Lives around slow-moving or still watercourses, including artificial dams, lakes and ornamental pools, where it feeds on invertebrates.

Male's call, a pulsed *creeeet creeeet creeeet*, repeated frequently, is made from grass tussocks, reeds, low, overhanging branches and emergent vegetation, from autumn to early spring. Eggs are laid in small, floating groups, sometimes attached to plant material, with about 50 eggs in a clutch, spread over a dozen or more floating groups. Tadpoles take about 18 weeks to change into frogs.

Aaron Payne

Long-snouted Frog ■ *Litoria longirostris* TL 26mm

DESCRIPTION Colouration varies from yellow to dark brown above, usually mottled with browns, greys and yellows with black spotting. Back smooth with raised tubercles, and snout pointed and elongated. Underparts whitish, with darker flecking. **DISTRIBUTION** Occurs in McIllwraith Range, Mt Tozer and Sir William Thomson range, northeastern Qld.

HABITS AND HABITAT Typically found in rainforests and vine thickets along waterways and seepages, where it feeds on invertebrates. Male's call has two parts – a short chirp, followed by deeper clicking – and is given from trees and shrubs surrounding water. It has a unique place for depositing its eggs in Australia, laying lay them on a plant overhanging the water. Average clutch size is 45 eggs.

Anders Zimny

Armoured Frog ■ *Litoria lorica* TL 43mm

DESCRIPTION Colouration varies from grey to grey-brown above, heavily mottled with greens, browns, greys and yellows, with black spotting. Colour and pattern resemble the granite boulders and plants of its natural habitat. Back rough and granular. Underside whitish. **DISTRIBUTION** Found in Carbine Tableland area of northeastern Qld. One of Australia's rarest frogs, having suffered massive decline due to chytrid fungus. Thought

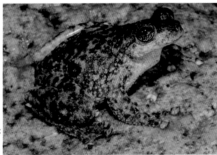

to be extinct until 2008, when a population was rediscovered west of previously recorded range. **HABITS AND HABITAT** Inhabits splash zones of rapidly moving waters and surrounding bushland. Call is unknown – a short low growl has been heard with the frogs in proximity, but cannot be presumed to be from this species. Reproductive biology poorly known, but thought to be similar to that of the Torrent Tree Frog (see p. 138).

Rockhole Frog ■ *Litoria meiriana* TL 23mm

DESCRIPTION Mottled grey or brown above, depending on colour of rock it is found on, and limbs banded in an irregular manner. Some individuals are reddish-tan with a yellow dorsolateral stripe. Small, raised tubercles across body help to disrupt the frog's shape, aiding camouflage. Underparts white to cream. Very similar to newly described Kimberley Rockhole Frog (see p. 114), but differs by its rounded snout and different call. **DISTRIBUTION** Occurs from Arnhem Plateau of northern NT, across and into

Kimberley region, WA. **HABITS AND HABITAT** Restricted to rocky gorges that have both permanent and season-dependent watercourses. Diminutive species that is different from most Australian frogs in that it is diurnal. Its small, agile body allows it to escape predators. It is thought to eat invertebrates. Its call sounds like a rapidly repeated *pop-pop-pop-pop-pop*. Produces about 60 eggs in a clutch. Tadpoles take about 4 weeks to become frogs.

Javelin Frog ■ *Litoria microbelos* TL 16mm

DESCRIPTION Colouration varies from light yellow to dark brown above, with white to cream flanks, and dark stripe that extends from snout, along side to just above front legs. Some individuals flecked with black. Back smooth, without bumps or protrusions. **DISTRIBUTION** Found in northeastern Australia, from Bluewater to Weipa in Qld, and across NT to Kimberley region of WA. **HABITS AND HABITAT** Occupies wide range of

habitats, including artificial and natural watercourses such as dams, melaleuca swamps, ponds and roadside ditches, where it feeds on invertebrates. Male's call is made while sitting on emergent plants, including reeds, or on water lilies. Call, intensified after rain, is a high-pitched *reeeee-reeeeee-reeeeeee*. Spawn is a series of small clusters, each consisting of about 12 eggs, which are attached to an emergent stem of a water plant. Can produce up to 889 eggs in a clutch.

Motorbike Frog ■ *Litoria moorei* TL 85mm
(Moore's Frog)

DESCRIPTION Body elongated. Colouration varies, but usually gold with large green blotches and spots, but can be almost all gold or all green in some individuals. Colouration can change depending on how warm the frog is; it is brightest when at its preferred body temperature. Back has a warty-like texture, with bumps or protrusions sometimes forming short ridges. **DISTRIBUTION** Found in southwestern corner of WA. **HABITS AND HABITAT** Favours swamps, large dams or lagoons, but will utilize areas surrounding

suburban garden ponds. Male's call is made from floating vegetation and among reeds near the water level, mainly after rain, from spring to mid-summer. Call has been described as sounding like a motorbike changing gears. Amplexus generally occurs after rain. Spawn is a floating mass usually attached to emergent vegetation. Tadpoles generally take 9–12 weeks to reach metamorphosis, but some individuals can take 60 weeks.

Scott Eipper

Scott Eipper

Kuranda Tree Frog
■ *Litoria myola* TL 72mm

DESCRIPTION Colouration varies from reddish-brown to grey-brown, heavily mottled with greens, browns, greys and yellows with black flecking, resembling mosses and lichens of a rainforest tree trunk. Legs have broad bands, and edges have tubercles to assist in breaking up the body shape of the frog. Back slightly rough and granular. Underside whitish. Upper iris turquoise. **DISTRIBUTION** Occurs around Kuranda in northeastern Qld. **HABITS AND HABITAT** Usually found in rainforests and vine thickets along waterways and seepages, where it feeds on invertebrates. By day it sits against trees, utilizing its excellent camouflage. Male's call is given from trees and rocks surrounding water. Call is a series of rapid clicks that sounds similar to that of the Green-eyed Tree Frog (see p. 151), but is much faster. Average clutch size is 509 eggs.

Torrent Tree Frog ■ *Litoria nannotis* TL 70mm
(Waterfall Frog)

DESCRIPTION Colouration varies from grey to grey-brown. Heavily mottled with greens, browns, greys and yellows, with black spotting, resembling boulders and plants of its natural habitat. Back rough and granular. Underside whitish. **DISTRIBUTION** Occurs

Scott Eipper

in wet tropics area of northeastern Qld. Has suffered serious decline due to chytrid fungus. **HABITS AND HABITAT** Found in splash zones of rapidly moving waters and surrounding bushland, where it feeds on invertebrates. Utilizes an interesting anti-predator strategy, involving leaping into the water and using the strong current to take it away from potential danger. Male's call is quite soft, sounding like *crok-crok-crok*, and is given from trees and shrubs surrounding water, as well as from exposed boulder faces. Average clutch is 160 eggs.

Rocket Frog ■ *Litoria nasuta* TL 65mm

DESCRIPTION Colouration a mix of ragged-edged stripes of browns, reds, yellows, greys and whites, often with faint flecking. Dark stripe edged with white or yellow extends from snout, along sides to just above front legs. Back covered in bumps or protrusions, aligned in longitudinal rows. Underside whitish. **DISTRIBUTION** Occurs along coast of eastern and northern Australia, from Port Stephens, NSW, to Kimberley region, WA. **HABITS AND HABITAT** Found in wallum, swamps, floodplains, savannah and open forests, where it feeds on invertebrates. Male's call, repeated short *clucks*, is made while sitting around still

waterbodies or on exposed points in a waterbody. The longest jumping frog in Australia – when fleeing predators it often jumps 2–3 times, changing direction each time. Spawn is laid across the water's surface, floating before eventually sinking to the bottom, with about 75 eggs in a clutch. Tadpoles develop into frogs in about 4 weeks.

Scott Eipper

Bridle Frog ■ *Litoria nigrofrenata* TL 50mm
(Tawny Rocket Frog)

DESCRIPTION Reddish-tan to yellowish above, with prominent dark stripe, edged with white or yellow, extending from snout, along sides to midway down body, with a break above the arm. Often a dark spot preceding groin on flanks. Back smooth. Underside whitish. **DISTRIBUTION** Found in Qld, from Cairns north, including Torres Strait Islands. Also occurs in southern PNG. **HABITS AND HABITAT** Inhabits swamps,

floodplains, savannah and open forests. Feeds on invertebrates. Male's call is made while sitting around still waterbodies or on exposed points in a waterbody. When fleeing predators it often jumps 2–3 times, changing direction each time. Call is a repeated, short *honk*. Spawn is laid across the water's surface, floating before eventually sinking to the bottom, and producing about 400 eggs in a clutch. Tadpoles develop into frogs in about 7 weeks.

Scott Eipper

Narrow-fringed Frog ▪ *Litoria nudidigita* TL 35mm
(Southern Leaf Green River Frog)

DESCRIPTION Bright green above, with dark-edged gold or yellow stripe extending from nostril above eye and broadening over flanks. Back smooth, without bumps or protrusions. Underside white. **DISTRIBUTION** Found in southeastern Australia, from just south of Sydney, NSW, to Gippsland, northeastern Vic. **HABITS AND HABITAT** Lives around creeks and streams in forests, breeding in ephemeral pools beside creeks and streams, and in

slower parts of the main waterway. Feeds on invertebrates. Male's call is made from overhanging vegetation or branches, usually about 30cm above the water level. Call is made in two parts – the first a *wrekik-wrekik* and the second a loud *eeep*, which is not always repeated. Breeds in ephemeral pools beside streams. Eggs are laid in loose clusters attached to sticks or plant material. Average clutch size is 430 eggs. Tadpoles transform to frogs after about 9 weeks.

Mountain Mist Frog ▪ *Litoria nyakalensis* TL 42mm

DESCRIPTION Olive-green, brown or grey above, with darker flecking, which is more prominent in some individuals. Lower flanks can be purplish. Back granular. Underside whitish. Similar to the Common Mist Frog (see p. 148), but has thicker forearms, larger thumb spines and a different call. **DISTRIBUTION** Formerly found at 380–1,020m asl from Thornton Peak to Cardwell in north Qld. Suffered massive decline due to chytrid fungus, most likely becoming extinct in about 1990. There is hope that, as is the case with

the Armoured Frog (see p. 136), it will be rediscovered. **HABITS AND HABITAT** Lived around creeks and streams in rainforests. Feeds on invertebrates. Male's call, a slow *creeeeek*, is made from overhanging vegetation or branches above the water level. Eggs are laid beneath rocks in riffle zones, with an average clutch size of 90 eggs. Duration as a tadpole is unknown.

Wallum Sedge Frog ■ *Litoria olongburensis* TL 29mm
(Olongburra, Sharp-snouted Reed Frog)

DESCRIPTION Colouration varies from light to dark green, with white to cream flanks. Dark stripe extends from snout, along sides to just above front legs. Some individuals flecked with black. Back smooth, without bumps or protrusions. Lower flanks can have orange, red and blue highlights. Nose distinctly pointed in profile. Underside white to cream.
DISTRIBUTION Found in eastern Australia, from Fraser Island, Qld, down to about Coffs Harbour, NSW. **HABITS AND**

HABITAT Occupies wallum swamps, ephemeral pools, ponds beside creeks, roadside ditches and adjacent grassland, where it feeds on invertebrates. One of the 'acid frogs', so named due to their ability to live in water with a low pH. Male's call, a *chip-chip* with an upwards pitching bobble, is made while sitting on emergent plants, including reeds, or on water lilies. Breeding biology unknown but thought to be similar to that of the Dwarf Tree Frog (see p. 129).

Pale Rocket Frog ■ *Litoria pallida* TL 41mm

DESCRIPTION Colouration highly variable. Usually plain pale brown to dark brown or grey. Some individuals bright yellow, others mottled with or without faint to even prominent blotching. Dark stripe extends from snout, along sides to just above front legs; conspicuous break in stripe before eyes. Back smooth. Underside whitish. **DISTRIBUTION** Occurs over most of northern Australia, from Townsville, Qld, to Kimberley region, WA.

HABITS AND HABITAT Found in swamps, floodplains, savannah and open forests, and also utilizes artificial waterbodies such as dams, where it feeds on invertebrates. Male's call is a short *wirr* and occasionally a repeated *chuk-chuk-chuck*, given while sitting around still waterbodies or on exposed points in a waterbody. Spawn is laid across the base of a waterbody, producing about 170 eggs, and tadpoles develop into frogs in about 8 weeks.

Victorian Frog ■ *Litoria paraewingi* TL 42mm

DESCRIPTION Colouration varies from light grey to brown, with dark brown to tan stripe that extends from snout through eye and along flanks. Some individuals can have additional stripes and dark flecks. Insides of hip and groin light orange without black markings. Back smooth, without bumps or protrusions. **DISTRIBUTION** Found in Vic and bordering Riverina region of NSW. **HABITS AND HABITAT** Lives around slow-moving or still watercourses, including artificial dams, lakes and ornamental pools, in grassland and open forests north of Great Dividing Range. Visually identical to the Brown Tree Frog (see

p. 128), but the call is different, as is the distribution. Feeds on invertebrates. Male's call, a drawn-out *creeee-creeee-creeee*, repeated quite frequently, is given from grass tussocks, low, overhanging branches and emergent vegetation, or while floating in the frigid water. Heard from autumn to early spring. Eggs are laid in small, floating groups, sometimes attached to plant material, with a clutch size of about 150 eggs. Tadpoles develop into frogs in about 25 weeks.

Cascade Tree Frog ■ *Litoria pearsoniana* TL 38mm

DESCRIPTION Bright green above, with dark-edged gold or yellow stripe extending from nostrils above eyes and broadening over flanks. Area below stripe often brown to light purple. Back smooth, without bumps or protrusions. Some individuals are mottled in appearances, looking similar to the Peppered Frog (see p. 144). Underside white. **DISTRIBUTION** Found from near Gympie, Qld, south to Gibraltar Range, NSW. Has suffered serious decline due to chytrid fungus, and is possibly a composite of up to three undescribed species. **HABITS AND HABITAT** Lives around creeks and streams in rainforests and wet sclerophyll forests, where it breeds in ephemeral pools beside creeks

and streams, and in slower parts of the main waterway. Feeds on invertebrates. Male's call is made from overhanging vegetation or branches, usually about 30cm above the water level. Call is in two parts – the first a *wrekik-wrekik* and the second a loud *eeep*, which is not always repeated. Eggs are laid in loose clusters, attached to sticks or plant material, with an average clutch size of 450 eggs. Tadpoles transform into frogs about 10 weeks after hatching.

Peron's Tree Frog ■ *Litoria peronii* TL 65mm
(Emerald Spotted Frog)

DESCRIPTION Colouration variable, but typically tan with small emerald green dots and spots during the day, and dark brown with black speckling at night, giving a mottled appearance. In some individuals green flecks become more easily seen at night. Webbing on hands, and insides of thighs and backs of legs, bright yellow and black. Back looks almost rough, with tiny, wart-like bumps or protrusions. **DISTRIBUTION** Occurs in eastern Australia, from Maryborough, Qld, to eastern SA. **HABITS AND HABITAT**
Found in swamps, large dams or lagoons, where it feeds on invertebrates. Male's call consists of series of closely spaced *arcs*, slowing towards the end of the call and resembling a twisted laugh, and is given from floating vegetation, on emergent and surrounding plants. Small groups of spawn are laid near the water's edge in leaf litter, or occasionally attached to submerged horizontal stems. Average clutch consists of about 1,750 eggs, and tadpoles take up to 14 weeks to complete metamorphosis (occasionally longer).

Scott Eipper

Masked Frog ■ *Litoria personata* TL 32mm

DESCRIPTION Colouration highly variable. Usually plain, pale brown to grey, although some individuals have bright yellow flanks, while others are mottled with faint blotching. Dark stripe extends from snout, along sides to just above front legs. Back smooth. Underparts whitish.
DISTRIBUTION Restricted to sandstone gorges and hills of western Arnhem region, NT. **HABITS AND HABITAT** Inhabits gorges and creeks in sandstone hills, utilizing pools of rainwater and springs, but also found among sedges and reeds along permanent water. Feeds on invertebrates. Male's call, a repeated *chirp* sounding almost electrical in origin, is made while perched on rock faces. Reproductive biology poorly known.

Angus McNab

Leaf-green Tree Frog ■ *Litoria phyllochroa* TL 40mm
(Green Stream Frog)

DESCRIPTION Bright green to dark brown above, with dark-edged gold or yellow stripe extending from nostrils, above eyes to armpits. Back smooth, without bumps or protrusions. Underside white. **DISTRIBUTION** Found in eastern NSW, from the Budderoo Plateau to near Bellingen. **HABITS AND HABITAT** Lives around creeks and streams in forests, breeding in ephemeral pools beside creeks and streams, and in slower parts of the main waterway. Feeds on invertebrates. Male's call, a loud *eeech*, repeated quite frequently, is

made from overhanging vegetation or branches, usually about 30cm above the water level, especially after heavy rain, from spring to early summer. Breeds in ephemeral pools beside streams, and eggs are laid in loose clusters, attached to sticks or plant material. A clutch size of 400 eggs has been recorded. Tadpoles start to develop into metamorphlings after 9 weeks.

Peppered Frog ■ *Litoria piperata* TL 32mm

DESCRIPTION Mottled dark green and grey above, and white below. Skin granular. Toe pads oval shaped, and webbing between toes and fingers greatly reduced. **DISTRIBUTION** Found historically at headwaters of Clarence River drainage in NSW. **HABITS AND HABITAT** Formerly found

around creeks and streams in forests with large boulders and thick vegetation, feeding on invertebrates. Call is described as *chuck-chuck-chuck*. May be extinct, with no confirmed sightings since 1973 despite intensive searching. Possibly a synonym of the Cascade Tree Frog (see p. 142). Further research is required to resolve the validity of this taxa.

Screaming Tree Frog ■ *Litoria quiritatus* TL 41mm

DESCRIPTION Cream to dark brown, with darker irregular stripe running from nose over shoulders down to near hips. Remaining areas usually light brown to grey, lightening on to lower flanks. Underside cream to white. Back smooth, without bumps or protrusions. Differs from the Slender Bleating Frog (see p. 115) by dorsolateral line posterior to forelimb and more robust body form. Differs from the Bleating Tree Frog (see p. 127) by yellow v dark vocal sac and more southerly distribution. **DISTRIBUTION** Found along Great Dividing Range from Giro, NSW, to Mallacoota, VIC. **HABITS AND HABITAT** Occurs in moist habitats such as swamps, rainforests, eucalypt forests, heaths and wallum. Also utilizes urban environments, including toilets, roadside ditches, cattle troughs and dams. Feeds on invertebrates. Male's call made from vegetation and bases of grass clumps around a waterbody. Call long and sounds like a bleating sheep. Lays an average clutch of 1,100 eggs, which once hatched, take about 8 weeks to turn into juvenile frogs.

Scott Eipper

Riverina Bell Frog *Litoria raniformis raniformis* TL 104mm

DESCRIPTION Colouration varies, but is usually gold with large green blotches and spots, and in some individuals almost all gold or all green. Depending on the frog's warmth, colouration can change, being brightest when it is at its preferred body temperature. Back has wart-like bumps or protrusions. The two subspecies of *L. raniformis* are separated by genetic divergence and distribution. **DISTRIBUTION** Found in southeastern Australia, along the Murray river region of SA, northern half of Vic and southern Riverina region of NSW. Introduced to NZ. **HABITAT AND HABITS** Prefers swamps, large dams or lagoons and other slow-moving or still bodies of water. Male's call is made from among floating

vegetation and on reeds near the water level, primarily after rain, from spring to autumn. Call has been described as a single, drawn-out *crawark-crawark-crok-crok*. Pond or pool breeder, preferring little or no surface movement. Spawn occurs in floating clusters but may also be submerged and wrapped around aquatic plants. Lays about 1,800 eggs in a clutch. Tadpoles take an average of 12 weeks to complete metamorphosis.

Southern Bell Frog *Litoria raniformis major* TL 104mm
(Growling Grass Frog)

DESCRIPTION Colouration varies, but is usually gold with large green blotches and spots, and in some individuals almost all gold or all green. Depending on the frog's warmth, colouration can change, being brightest when it is at its preferred body temperature. Back has wart-like bumps or protrusions. The two subspecies of *L. raniformis* are separated by genetic divergence and distribution. **DISTRIBUTION** Found in SA across the Naracoorte plain through southern Vic with a record in south-east NSW near Bombala. Found across Tas. **HABITAT AND HABITS** Prefer swamps, large dams or lagoons and other slow-

moving or still bodies of water. Male's call is made from among floating vegetation and on reeds near the water level, primarily after rain, from spring to autumn. Call has been described as a single, drawn-out *crawark-crawark-crok-crok*. Pond or pool breeder, preferring little or no surface movement. Spawn occurs in floating clusters but may also be submerged and wrapped around aquatic plants. Lays about 2,400 eggs in a clutch. Tadpoles take an average of 15 weeks to complete metamorphosis.

Whirring Tree Frog ■ *Litoria revelata* TL 36mm
(Revealed Frog)

DESCRIPTION Colouration varies from light grey to yellow, with or without dark brown to black stripe that extends from snout through eye and along flanks. Some individuals can have additional stripes and dark flecks. Insides of hips and groin light orange with black markings. Back smooth, without bumps or protrusions. **DISTRIBUTION** Found in three disjunct populations from Ourimbah, NSW, to Qld border, around Eungella, Qld, and Wet Tropics of Qld. Possibly a species complex; northern population is named *Litoria corbeni* – further work is needed to determine validity of this taxa. **HABITS AND HABITAT** Lives around slow-moving or still watercourses, including artificial dams and lakes. Feeds on invertebrates. Male's call is made from grass tussocks, low, overhanging branches and emergent vegetation, from autumn to early spring. Call is a drawn-out *reek-reek-reek*, repeated frequently. Can have a clutch size of about 500 eggs attached to vegetation. Tadpoles take an average of 12 weeks to complete metamorphosis.

Scott Eipper

Aaron Payne

Northern population

Common Mist Frog ■ *Litoria rheocola* TL 43mm
(Creek Frog)

DESCRIPTION Olive-green, brown or grey above, with darker flecking that is more prominent in some individuals. Lower flanks can be purplish, sometimes with pale, crescent-shaped markings on flanks, and underparts are whitish. Back granular. Similar to the Mountain Mist Frog (see p. 140), but has thinner forearms, smaller thumb spines and a different call. **DISTRIBUTION** Found from Rossville to Ingham, northern Qld. Has suffered significant declines due to chytrid fungus, leading to disappearance from upland sites. **HABITS AND HABITAT** Lives around creeks and streams in rainforests, where it feeds on invertebrates. Male's call, a slow *creeeek*, is made from overhanging vegetation or branches above the water level. About 45–60 white eggs are laid beneath rocks in riffle zones. Duration as a tadpole unknown.

Scott Eipper

Roth's Tree Frog ■ *Litoria rothii* TL 58mm

DESCRIPTION Colouration varies, but usually consists of cream to tan during the day, and dark brown with black speckling, giving a mottled appearance, at night. Webbing on hands, insides of thighs and rear of legs bright yellow and black. Upper of eye bright red. Back looks rough, with tiny, wart-like bumps. **DISTRIBUTION** Found from Maryborough, Qld, across northern Australia to Kimberley region, WA. **HABITS AND HABITAT** Prefers swamps, large dams or lagoons, where it feeds on invertebrates. Male's call, a series of closely spaced *arcs*, slowing towards the end of the call, is given from floating vegetation, emergent plants or shrubs surrounding a waterbody. Spawn is small groups laid near the water's edge, with an average clutch size of about 504 eggs. Tadpoles take about 9–16 weeks to complete metamorphosis.

Daytime colouration

Desert Tree Frog ■ *Litoria rubella* TL 43mm
(Red Tree Frog, Naked Tree Frog)

DESCRIPTION Colouration varies from cream to dark brown, with dark brown to black stripe that runs along length of body. Bottom edge of stripe often merges with side colouration on to white ventral surface. Back smooth, without bumps or protrusions. Some individuals have pale vertebral stripe. **DISTRIBUTION** Occurs across most of Australia, from mid-NSW inland to WA, and also PNG. Probably a species complex, some individuals from northern NT resembling *L. congenita* from PNG. **HABITS AND HABITAT** Found in dry habitats such as deserts, brigalow and savannah, where it seeks out both natural and artificial waterbodies, including dams and cattle troughs, and even shelters in toilets. Feeds on invertebrates. Male's call is long and distinctly pulsed, slowing and rising towards the end, and is given from the ground around a waterbody, after rain, from summer to autumn. Spawn are small clusters laid near the water's edge in temporary pools, with an average clutch containing 200 eggs. Once hatched, tadpoles take about 5 weeks to turn into juvenile frogs.

Green-eyed Tree Frog ■ *Litoria serrata* TL 80mm

DESCRIPTION Colouration varies from reddish-brown to grey-brown, heavily mottled with greens, browns, greys and yellows with black flecking, resembling mosses and lichens of the rainforest. Legs have broad bands, and edges of legs have tubercles to assist in breaking up the body shape of the frog. Back slightly rough and granular. Underparts whitish. Upper iris turquoise. **DISTRIBUTION** Ranges from Mt Spec to Cooktown, Qld. **HABITS AND HABITAT** Usually found in rainforests and vine thickets along waterways and seepages, where it feeds on invertebrates. During the day sits against trees, relying on its excellent camouflage for protection. Male's call consists of a series of clicks, and is given from trees and rocks surrounding water. Average clutch comprises 740 eggs, and tadpoles transform to frogs about 9 weeks after hatching.

Scott Eipper

Spotted Tree Frog ■ *Litoria spenceri* TL 51mm

DESCRIPTION Light tan to dark brown or grey above, although some individuals are bright green. Can be mottled or spotted, and sometimes has a gold stripe that extends along canthal ridge. Back has bumps or protrusions. Cream to white below. **DISTRIBUTION** Occurs in NSW and Vic high country. Has suffered major declines due to chytrid fungus, trout introduction and habitat modification, and is now found in just a few locations. **HABITS AND HABITAT** Found along natural watercourses, such as rocky streams, river headwaters and creeks, where it is often seen on rocks and among adjacent vegetation around the water's edge. Feeds on invertebrates. Male's call is made in two parts – the first a *reeee* and the second a repeated whirring – and is given from vegetation near water. Stream breeder, with spawn comprising a mass of about 525 eggs, adhered to gravel beneath rocks in a shallow section of a stream. After hatching, tadpoles take about 12 weeks to become frogs.

Adam Elliott

Northern Creek Frog ■ *Litoria spaldingi* TL 70mm

DESCRIPTION Reddish-tan to yellowish or dark brown above, with prominent dark stripe that extends from snout, along side and down flanks. Back smooth. Underparts whitish. Not distinguishable morphologically from the Wotjulum Frog (see p. 156) but tends to have fewer yellow markings on posterior of thigh. **DISTRIBUTION** Occurs in northern Australia from Selwyn, QLD, to Litchfield National Park, NT. Distribution does not overlap with Wotjulum Frog's.

HABITS AND HABITAT Found in rocky gorges and surrounding open forests near permanent water, where it feeds on invertebrates. Male's call very variable, with short chucks, clicks and whirring, and given while partially obscured by vegetation near a waterbody. Spawn laid in floating clumps of up to 300 eggs, and tadpoles develop into frogs about 7 weeks after hatching.

Magnificent Tree Frog ■ *Litoria splendida* TL 118mm
(Splendid Tree Frog)

DESCRIPTION Robust frog. Colouration varies from light to dark green, with stunning yellow spots and flecks. Back smooth, without bumps or protrusions. Pair of large glands behind head help distinguish it from the similar Green Tree Frog (see p. 119).

Underparts cream to white. **DISTRIBUTION** Occurs in northern WA and into NT border region. **HABITS AND HABITAT** Found around slow-moving or still water, as well as in artificial structures that hold water, such as buildings. Male's call, a deep *brawwwk*, has been recorded from 1.5–2m above the water after summer rains. Around 7,000 eggs are laid in a series of floating clumps, which then sink and hatch shortly after. Tadpoles take about 9 weeks to become frogs.

Chattering Rock Frog ■ *Litoria staccato* TL 41mm

DESCRIPTION Mottled grey or brown above, depending on the rock colour it is found on, and flanks usually lighter brown or pinkish. Body finely granular and depressed. Usually a dark streak from nose to midway down sides, but it may be faint. Underside white to cream. Much less webbing between toes compared to Copland's Rock Frog (see p. 122). **DISTRIBUTION** Found in Kimberley region, WA. **HABITS AND HABITAT** Restricted

to rocky gorges and environments that have both permanent and season-dependent watercourses, and the depressed body is a probable adaptation for sheltering in rock crevices. Thought to eat invertebrates. Male's call is a series of irregularly spaced, rapid chirps, given while sitting on rocks and branches in and around a waterbody. Produces about 30 eggs, laid on to small rocks in pools, and tadpoles take about 7 weeks to become frogs.

Adam Elliott

Glandular Frog ■ *Litoria subglandulosa* TL 60mm

DESCRIPTION Colouration variable, but usually brown above, with green and white to cream on flanks, becoming orange to pink on lower flanks and mottled black and white towards rear. Dark stripe edged with white-cream extends from snout, along sides to just above front legs. Below is green. Some individuals flecked with black or dark brown. Underside white. Back smooth, without bumps or protrusions. **DISTRIBUTION** Occurs in northeastern NSW, from Walcha to Stanthorpe, Qld. **HABITS AND HABITAT** Lives

around creeks and streams in forests and montane heaths. Has suffered significant declines due to chytrid fungus and habitat loss. Feeds on invertebrates. Male's call, which sounds like a *waahh rarr*, repeated every second or so, is given from the water's edge, on rocks in the water, or from overhanging vegetation. Calls are heard in spring to early summer. Breeds in ephemeral pools beside streams, and can lay an average clutch of 400 eggs. Tadpoles transform into frogs 15 weeks after hatching.

Scott Eipper

Tornier's Frog ■ *Litoria tornieri* TL 35mm

DESCRIPTION Reddish-tan to yellowish above, with prominent dark stripe that extends from snout, along side, to above arm. Back smooth. Underparts whitish. **DISTRIBUTION** Occurs in northern Australia, from Gulf Country of NT/Qld border to Kimberley region,

WA. **HABITS AND HABITAT** Inhabits swamps, floodplains, savannah and open forests, where it feeds on invertebrates. Male's call, a short *wack*, repeated quickly, is made while partially obscured by vegetation near a waterbody. Spawn is laid across the base of the waterbody, and contains about 350 eggs. Tadpoles develop into frogs about 7 weeks after hatching.

Tyler's Tree Frog ■ *Litoria tyleri* TL 52mm

DESCRIPTION Colouration variable, but typically tan with small emerald green dots and spots, but can change to dark brown with black speckling, giving a mottled appearance. In some individuals green flecks become more easy to see at night. Back rough, with tiny, wart-like bumps or protrusions. Webbing of hands, insides of thighs and backs of legs yellow and dark brown. **DISTRIBUTION** Found in eastern Australia, from Childers,

Qld, to Batemans Bay, NSW. **HABITS AND HABITAT** Prefers swamps, large dams or lagoons, where it feeds on invertebrates. Male's call, a series of closely spaced *arcs*, is made from floating vegetation, emergent plants, and among trees and shrubs surrounding a waterbody. Around 450 eggs are laid in a clutch, near the edge of the water, and tadpoles take about 10 weeks to complete metamorphosis after hatching.

Alpine Tree Frog ■ *Litoria verreauxii alpina* TL 35mm

DESCRIPTION Typically light grey to brown above, with dark brown to tan stripe that extends from snout, through eye and along flanks. Some individuals striped green and brown. Insides of hips and groin light orange, almost always with black markings. Back smooth, without bumps or protrusions. **DISTRIBUTION** Found from Vic to NSW and ACT in Southern Alps. Has suffered major declines due to chytrid fungus and modification of habitat, including breeding sites, by feral species. **HABITS AND HABITAT** Lives around slow-moving or still watercourses, where it feeds on invertebrates. Male's call, a rapid *cree cree cree cree*, repeated often, is made from grass tussocks, low, overhanging branches and emergent vegetation. Around 100 eggs are laid in a clutch, spread over a dozen or more floating groups, and tadpoles take about 15 weeks to complete metamorphosis.

Adam Elliott

Whistling Tree Frog ■ *Litoria verreauxii verreauxii* TL 36mm
(Verreaux's Frog)

DESCRIPTION Colouration varies from light grey to brown, with dark brown to tan stripe that extends from snout, through eye and along flanks. Some individuals can have additional stripes and dark flecks. Insides of hips and groin light orange and almost always have black markings. Back smooth, without bumps or protrusions. **DISTRIBUTION** Found from Vic to south-east Qld. **HABITS AND HABITAT** Occurs around slow-moving or still watercourses, where it feeds on invertebrates. Male's call, a pulsed *creeee creeee creeee* that is repeated often, is made from grass tussocks, low, overhanging branches and emergent vegetation. Eggs laid in small, floating groups, sometimes attached to plant material, with a clutch of about 720 eggs spread over a dozen or more floating groups. Tadpoles take about 13 weeks to complete metamorphosis.

Scott Eipper

Wotjulum Frog ■ *Litoria watjulumensis* TL 71mm

DESCRIPTION Reddish-tan to yellowish or dark brown above, with prominent dark stripe that extends from snout, along sides and down flanks. Back smooth. Underparts whitish. Not distinguishable morphologically from *L. spaldingi* (see p. 152) but tends to have a greater presence of yellow markings on posterior of thigh. **DISTRIBUTION** Occurs in northern Australia, from the Victoria River region, NT to the Kimbolton Spring in the

Kimberleys, WA. **HABITS AND HABITAT** Found in rocky gorges and surrounding open forests near permanent water, where it feeds on invertebrates. Male's call is very variable, with short chucks, clicks and whirring, and is given while partially obscured by vegetation near a waterbody. Spawn is laid in floating clumps of up to 300 eggs, and tadpoles develop into frogs about 7 weeks after hatching.

Southern Heath Frog ■ *Litoria watsoni* TL 70mm

DESCRIPTION Light grey to dark brown, with or without broad dark brown stripe that extends from behind eyes over head and down back. Some individuals can have additional stripes and dark flecks. Insides of armpits and groin pale orange. Back smooth. Differs from the Northern Heath Frog (see p. 135) by having fewer call pulses (a mean of 22.8 v 27.8), different genetics and more southerly distribution. **DISTRIBUTION** Found from Bonang, VIC, to Gerringong Falls, NSW. Has suffered significant decline and range contraction due to chytrid fungus and habitat loss. **HABITS AND HABITAT** Lives around slow-moving or still watercourses, including artificial dams, lakes and ornamental pools, where it feeds on

invertebrates. Male's call, a pulsed *creeeet creeeet creeeet*, repeated often, made from grass tussocks, reeds, low, overhanging branches and emergent vegetation, from autumn to early spring. Eggs laid in small floating groups, sometimes attached to plant material, with about 50 eggs in a clutch, spread over a dozen or more floating groups. Tadpoles take about 18 weeks to change into frogs.

Wilcox's Frog ■ *Litoria wilcoxii* TL 71mm

DESCRIPTION Dark brown above, with dark stripe that extends from snout, through eye and along flanks, and backs of thighs spotted with light yellow. Male exhibits distinct colour change to bright yellow when in breeding mode. Back smooth, without bumps or protrusions. **DISTRIBUTION** Found in northeastern Australia, from Goulburn, NSW, to Mareeba, Qld. **HABITS AND HABITAT** Occurs around creeks and streams in forests, breeding in ephemeral pools beside creeks and streams, and in slower parts of the main waterway, and feeding on invertebrates. Call is a soft, purring *bobble*, repeated quite frequently. About 3,000 eggs are laid in a mass, loosely attached to the rocky base of a stream or on plant material, and tadpoles start to develop into metamorphlings about 8 weeks after hatching.

Scott Eipper

Orange-thighed Frog ■ *Litoria xanthomera* TL 62mm

DESCRIPTION Light to dark green above, with vivid orange on backs of thighs, and almost smooth back. Iris bright red to orange. Underparts yellow to orange. **DISTRIBUTION** Found across northeastern Qld, from Townsville north to Cooktown. **HABITS AND HABITAT** Seeks out small waterbodies, both natural and artificial, including roadside ditches and dams, and ephemeral pools beside rivers, streams and creeks, where it feeds on invertebrates. Male's call is in two parts – the first consists of a drawn-out *wahh wahh wahh wahh wahh*, and the second, which is not always repeated, is a soft trill – given from vegetation and on the ground. About 1,100 eggs are laid in a floating mass, and tadpoles start to develop into metamorphlings about 6 weeks after hatching.

Scott Eipper

> **RANIDAE (TRUE FROGS)**
> The true frogs are probably recent natural invaders to Australia via New Guinea.
> They comprise one of the most diverse frog families worldwide, and in Australia
> are represented by a single species that also occurs in New Guinea. They are strong
> jumpers that are primarily terrestrial in Australia. Due to recent taxonomic changes,
> Water Frogs have moved from the genus *Rana*, to *Hylarana* and now *Papurana*.

Water Frog ■ *Papurana daemeli* TL 80mm
(Wood Frog)

DESCRIPTION Yellow-brown to grey above, with or without spots and flecks, and
light barring on legs. Body has low bumps and strong dorsolateral ridge. White below,
with or without brown flecking. One of Australia's few frogs with paired vocal sacs.
DISTRIBUTION Occurs in northeastern Qld, from Townsville north, with isolated
population in northeastern NT. Also found in PNG. **HABITS AND HABITAT** Lives
around waterways in forests, breeding in ephemeral pools beside creeks and streams and in
slower parts of the main waterway. Feeds on invertebrates. Call is variable but usually has
two distinct note types, similar to *wrekk wah wah wah wah*. Up to 20,000 eggs are laid in a
floating jelly mass.

Scott Eipper

BUFONIDAE (TRUE TOADS)
True toads are widely distributed around the world, and there are two introduced species
in Australia. The Cane Toad was released at Gordonvale, Qld, in 1935 to control pests
of sugar cane, and has since become a major pest. The South-east Asian Toad has a
small number of widespread records.

South-east Asian Toad ▪ *Duttphrynus melanostictus* TL 80mm
(Black-spined Toad)

DESCRIPTION Brown or greyish above, with black spotting on ends of the numerous tubercles, and numerous elevated black ridges and spines on limbs and body. Pale brown to grey below with white or cream marbling. **DISTRIBUTION** Not currently established in Australia. Numerous feral populations have become a problem in locations such as West Papua, Bali (Indonesia) and Madagascar (Africa). Accidentally imported into Australia, with 110 individuals collected in Melbourne, Sydney, Perth, Darwin and Cairns, most of which have been within the boundaries of cargo-ship terminals and airports. **HABITS AND HABITAT** Mainly nocturnal. Lives in wide variety of habitats, thriving in urban areas. Could have a devastating effect on Australian ecosystems due to its toxic nature. Lays up to 40,000 black eggs in various long strands. Tadpoles start to develop into metamorphlings after about 5–13 weeks.

Angus McNab

Cane Toad ■ *Rhinella marina* TL 120mm (exceptions up to 200mm)
(Marine Toad)

DESCRIPTION Yellow to brown or greyish above with black spotting; younger individuals can have red to orange spots. Juveniles grey to black, often flecked with orange and brown. Belly pale brown to grey with white or cream marbling. **DISTRIBUTION** Constantly changing and expanding. Current range has expanded as far south as NSW north coast and as far west as Mitchell Plateau, WA. **HABITS AND HABITAT** Lives in wide variety of habitats, including mangroves, swamps, forests, grassland, savannah and rocky gorges, and also thrives in agricultural and urban areas. Mainly nocturnal as adults, while juveniles and

metamorphs are often seen around waterbodies during the day. Feeds mostly on invertebrates and small vertebrates. Has had a devastating effect on Australian ecosystems due to its toxic nature. Recent evidence shows that many species are beginning to develop resistance or strategies that allow them to eat toads. Lays up to 35,000 black eggs in various long strands, and tadpoles start to develop into metamorphlings after about 5 weeks.

Juvenile

Taxonomy follows Cogger 2022, with the exception of the addition of newly described taxa. There is a pending significant revision of the Australasian tree frogs that will change genus-level taxonomy for many of Australia's species. The table on p. 169 summarizes these changes.

For each species, an 'x' indicates presence in a particular state or territory. State or territory abbreviations are as follows:

Qld	Queensland
NSW	New South Wales (including ACT)
Vic	Victoria
Tas	Tasmania
SA	South Australia
WA	Western Australia
NT	Northern Territory

Abbreviations of IUCN Red List Status (or equivalent for new taxa):

EX	Extinct
CE	Critically Endangered
EN	Endangered
VU	Vulnerable
NT	Near Threatened
LC	Least Concern
DD	Data Deficient
NA	Not Assessed

Common English Name	Scientific Name	Qld	NSW	Vic	Tas	SA	WA	NT	IUCN
Swamp Frogs									
Tusked Frog	Adelotus brevis	x	x						LC
Western Spotted Frog	Heleioporus albopunctatus						x		LC
Giant Burrowing Frog	Heleioporus australiacus australiacus		x						EN
Southern Owl Frog	Heleioporus australiacus flavopunctatus		x	x					EN
Hooting Frog	Heleioporus barycragus						x		LC
Moaning Frog	Heleioporus eyrei						x		LC
Plain Frog	Heleioporus inornatus						x		LC
Sand Frog	Heleioporus psammophilus						x		LC
Sandpaper Frog	Lechriodus fletcheri	x	x						LC
Marbled Frog	Limnodynastes convexiusculus	x					x	x	LC
Flat-headed Frog	Limnodynastes depressus						x	x	LC
Western Banjo Frog	Limnodynastes dorsalis						x		LC
Eastern Banjo Frog	Limnodynastes dumerilii dumerilii	x	x	x		x			LC
Snowy Mountains Banjo Frog	Limnodynastes dumerilii fryi		x						LC
Coastal Banjo Frog	Limnodynastes dumerilii grayi		x						LC
Southern Banjo Frog	Limnodynastes dumerilii insularis		x	x	x				LC
Mottled Banjo Frog	Limnodynastes dumerilii variegata			x	x	x			LC

Common English Name	Scientific Name	Qld	NSW	Vic	Tas	SA	WA	NT	IUCN
Barking Marsh Frog	Limnodynastes fletcheri	x	x	x		x			LC
Giant Banjo Frog	Limnodynastes interioris		x	x					LC
Carpenter Frog	Limnodynastes lignarius						x	x	LC
Striped Marsh Frog	Limnodynastes peronii	x	x	x	x	x			LC
Salmon-striped Marsh Frog	Limnodynastes salmini	x	x						LC
Spotted Marsh Frog	Limnodynastes tasmaniensis	x	x	x	x	x			LC
Northern Banjo Frog	Limnodynastes terraereginae	x	x						LC
White-footed Frog	Neobatrachus albipes						x		LC
Northen Trilling Frog	Neobatrachus aquilonius						x	x	LC
Tawny Trilling Frog	Neobatrachus fulvus						x		LC
Kunapalari Frog	Neobatrachus kunapalari						x		LC
Humming Frog	Neobatrachus pelobatoides						x		LC
Painted Trilling Frog	Neobatrachus pictus			x		x			LC
Sudell's Trilling Frog	Neobatrachus sudellae	x	x	x		x	x	x	LC
Shoemaker Frog	Neobatrachus sutor						x	x	LC
Plonking Frog	Neobatrachus wilsmorei						x		LC
Crucifix Spadefoot	Notaden bennetti	x	x						LC
Northern Spadefoot	Notaden melanoscaphus	x					x	x	LC
Desert Spadefoot	Notaden nichollsi	x				x	x	x	LC
Weigel's Spadefoot	Notaden weigeli						x		LC
Baw Baw Frog	Philoria frosti			x					CE
Mount Ballow Mountain Frog	Philoria knowlesi	x	x						EN
Red and Yellow Mountain Frog	Philoria kundagungan	x	x						EN
Loveridge's Frog	Philoria loveridgei	x	x						EN
Pugh's Mountain Frog	Philoria pughi		x						EN
Richmond Range Sphagnum Frog	Philoria richmondensis		x						EN
Sphagnum Frog	Philoria sphagnicolus		x						EN
Ornate Burrowing Frog	Platyplectrum ornatum	x	x				x	x	LC
Spencer's Burrowing Frog	Platyplectrum spenceri	x				x	x	x	LC
Narrow-mouthed Frogs									
Northern Territory Frog	Austrochaperina adelphe							x	LC
Fry's Frog	Austrochaperina fryi	x							LC
Slender Whislting Frog	Austrochaperina gracilipes	x							LC
White-browed Whistling Frog	Austrochaperina pluvialis	x							LC
Robust Whistling Frog	Austrochaperina robusta	x							LC

Common English Name	Scientific Name	Qld	NSW	Vic	Tas	SA	WA	NT	IUCN
Tapping Nursery Frog	Cophixalus aenigma	x							VU
Southern Ornate Nursery Frog	Cophixalus australis	x							LC
Buzzing Nursery Frog	Cophixalus bombiens	x							NT
Beautiful Nursery Frog	Cophixalus concinnus	x							CE
McIlwraith Nursery Frog	Cophixalus crepitans	x							VU
Northern Tapping Nursery Frog	Cophixalus exiguus	x							NT
Hinchinbrook Island Nursery Frog	Cophixalus hinchinbrookensis	x							VU
Hosmer's Nursery Frog	Cophixalus hosmeri	x							NT
Creaking Nursery Frog	Cophixalus infacetus	x							LC
Kutini Boulder Frog	Cophixalus kulakula	x							NA
Mount Elliot Nursery Frog	Cophixalus mcdonaldi	x							VU
Mountain-top Nursery Frog	Cophixalus monticola	x							EN
Neglected Nursery Frog	Cophixalus neglectus	x							EN
Northern Ornate Nursery Frog	Cophixalus ornatus	x							LC
Golden-capped Nursery Frog	Cophixalus pakayakulangun	x							NA
Cape York Nursery Frog	Cophixalus peninsularis	x							VU
Blotched Boulder Frog	Cophixalus petrophilus	x							NA
Black Mountain Boulder Frog	Cophixalus saxatilis	x							NT
Cape Melville Boulder Frog	Cophixalus zweifeli	x							DD
Ground Frogs									
White-bellied Frog	Anstisia alba						x		CE
Walpole Frog	Anstisia lutea						x		LC
Karri Froglet	Anstisia rosea						x		LC
Yellow-bellied Frog	Anstisia vitellina						x		CE
Northern Sandhill Frog	Arenophryne rotunda						x		LC
Southern Sandhill Frog	Arenophryne xiphorhyncha						x		LC
Hip Pocket Frog	Assa darlingtoni	x	x						LC
Mount Wollumbin Hip Pocket Frog	Assa wollumbin		x						CE
Bilingual Froglet	Crinia bilingua						x	x	LC
Desert Froglet	Crinia deserticola	x	x			x	x		LC
Kimberley Froglet	Crinia fimbriata						x		NA
Northern Flinders Ranges Froglet	Crinia flindersensis					x			NA
Tschudi's Froglet	Crinia georgiana						x		LC
Clicking Froglet	Crinia glauerti						x		LC

Common English Name	Scientific Name	Qld	NSW	Vic	Tas	SA	WA	NT	IUCN
Squelching Froglet	Crinia insignifera						x		LC
Moss Froglet	Crinia nimba				x				LC
Eastern Sign-bearing Froglet	Crinia parinsignifera	x	x	x		x			LC
Bleating Froglet	Crinia pseudinsignifera						x		LC
Northern Froglet	Crinia remota	x					x	x	LC
Southern Flinders Ranges Froglet	Crinia riparia					x			LC
Common Froglet	Crinia signifera	x	x	x	x	x			LC
Sloane's Froglet	Crinia sloanei		x	x					DD
South Coast Froglet	Crinia subinsignifera						x		LC
Tasmanian Froglet	Crinia tasmaniensis				x				LC
Tinkling Froglet	Crinia tinnula	x	x						VU
Smooth Frog	Geocrinia laevis			x	x				LC
Lea's Frog	Geocrinia leai						x		LC
Victorian Smooth Frog	Geocrinia victoriana		x	x					LC
Forest Toadlet	Metacrinia nichollsi						x		LC
Stuttering Frog	Mixophyes balbus		x	x					EN
Carbine Barred Frog	Mixophyes carbinensis	x							LC
Mottled Barred Frog	Mixophyes coggeri	x							LC
Great Barred Frog	Mixophyes fasciolatus	x	x						LC
Fleay's Barred Frog	Mixophyes fleayi	x	x						EN
Giant Barred Frog	Mixophyes iteratus	x	x						VU
Northern Barred Frog	Mixophyes schevilli	x							LC
Turtle Frog	Myobatrachus gouldii						x		LC
Haswell's Frog	Paracrinia haswelli		x	x					LC
Red-crowned Broodfrog	Pseudophryne australis		x						VU
Bibron's Broodfrog	Pseudophryne bibroni	x	x	x		x			NT
Red-backed Broodfrog	Pseudophryne coriacea	x	x						LC
Southern Corroboree Frog	Pseudophryne corroboree		x						CE
Magnificent Broodfrog	Pseudophryne covacevichae	x							EN
Dendy's Broodfrog	Pseudophryne dendyi		x	x					LC
Gorge Broodfrog	Pseudophryne douglasi						x		LC
Crawling Broodfrog	Pseudophryne guentheri						x		LC
Large Broodfrog	Pseudophryne major	x	x						LC
Western Broodfrog	Pseudophryne occidentalis						x		LC
Northern Corroboree Frog	Pseudophryne pengilleyi		x						CE
Copper-backed Broodfrog	Pseudophryne raveni	x							LC

Common English Name	Scientific Name	Qld	NSW	Vic	Tas	SA	WA	NT	IUCN
Central Ranges Broodfrog	*Pseudophryne robinsoni*					x			NA
Southern Broodfrog	*Pseudophryne semimarmorata*			x	x				LC
Southern Gastric Brooding Frog	*Rheobatrachus silus*	x							EX
Northern Gastric Brooding Frog	*Rheobatrachus vitellinus*	x							EX
Sunset Frog	*Spicospina flammocaerulea*						x		VU
Sharp-nosed Day Frog	*Taudactylus acutirostris*	x							CE
Southern Day Frog	*Taudactylus diurnus*	x							EX
Eungella Day Frog	*Taudactylus eungellensis*	x							CE
Eungella Tinker Frog	*Taudactylus liemi*	x							NT
Kroombit Tinker Frog	*Taudactylus pleione*	x							CE
Northern Tinker Frog	*Taudactylus rheophilus*	x							CE
Montane Toadlet	*Uperoleia altissima*	x							LC
Jabiru Toadlet	*Uperoleia arenicola*							x	DD
Derby Toadlet	*Uperoleia aspera*						x		LC
Northern Toadlet	*Uperoleia borealis*						x	x	LC
Fat Toadlet	*Uperoleia crassa*	x					x	x	LC
Howard Springs Toadlet	*Uperolia daviesae*							x	DD
Dusky Toadlet	*Uperoleia fusca*	x	x						LC
Glandular Toadlet	*Uperoleia glandulosa*						x		LC
Gurrumul's Toadlet	*Uperoleia gurrumuli*							x	NA
Smooth Toadlet	*Uperoleia laevigata*	x	x	x					LC
Stonemason Toadlet	*Uperoleia lithomoda*	x					x	x	LC
Littlejohn's Toadlet	*Uperoleia littlejohni*	x							LC
Mahony's Toadlet	*Uperoleia mahonyi*		x						DD
Marbled Toadlet	*Uperoleia marmorata*						x		DD
Martin's Toadlet	*Uperoleia martini*			x	x				DD
Tiny Toadlet	*Uperoleia micra*						x		NA
Tanami Toadlet	*Uperoleia micromeles*						x	x	LC
Mimic Toadlet	*Uperoleia mimula*	x							LC
Small Toadlet	*Uperoleia minima*						x		LC
Mjoberg's Toadlet	*Uperoleia mjobergii*						x		LC
Alexandria Toadlet	*Uperoleia orientalis*							x	DD
Wrinkled Toadlet	*Uperoleia rugosa*	x	x	x		x			LC
Russell's Toadlet	*Uperoleia russelli*						x		LC
Pilbara Toadlet	*Uperoleia saxatilis*						x		NA
Ratcheting Toadlet	*Uperoleia stridera*						x	x	NA

Common English Name	Scientific Name	Qld	NSW	Vic	Tas	SA	WA	NT	IUCN
Mole Toadlet	Uperoleia talpa						x		LC
Blacksoil Toadlet	Uperoleia trachyderma	x						x	LC
Tyler's Toadlet	Uperoleia tyleri		x	x					DD
Australian Tree Frogs									
Striped Burrowing Frog	Cyclorana alboguttata	x	x			x			LC
Giant Frog	Cyclorana australis	x					x	x	LC
Short-footed Frog	Cyclorana brevipes	x	x						LC
Hidden-ear Frog	Cyclorana cryptotis	x					x	x	LC
Knife-footed Frog	Cyclorana cultripes	x					x	x	LC
Long-footed Frog	Cyclorana longipes	x					x	x	LC
Daly Waters Frog	Cyclorana maculosa	x					x	x	LC
Main's Frog	Cyclorana maini		x			x	x	x	LC
Small Frog	Cyclorana manya	x					x	x	LC
New Holland Frog	Cyclorana novaehollandiae	x	x			x			LC
Western Water-holding Frog	Cyclorana occidentalis						x		LC
Eastern Water-holding Frog	Cyclorana platycephala	x	x					x	LC
Wailing Frog	Cyclorana vagita						x	x	LC
Rough Frog	Cyclorana verrucosa	x	x						LC
Slender Tree Frog	Litoria adelaidensis						x		LC
Cape Melville Tree Frog	Litoria andiirrmalin	x							VU
Green and Golden Bell Frog	Litoria aurea		x	x					VU
Kimberley Rockhole Frog	Litoria aurifera						x		LC
Kimberley Rocket Frog	Litoria axillaris						x		LC
Slender Bleating Tree Frog	Litoria balatus	x							LC
Barrington Tops Tree Frog	Litoria barringtonensis		x						LC
Beautiful Tree Frog	Litoria bella	x							LC
Northern Dwarf Tree Frog	Litoria bicolor	x					x	x	LC
Booroolong Frog	Litoria booroolongensis		x	x					CE
Green-thighed Frog	Litoria brevipalmata	x	x						EN
Tasmanian Tree Frog	Litoria burrowsae				x				LC
Green Tree Frog	Litoria caerulea	x	x			x	x	x	LC
Yellow-spotted Tree Frog	Litoria castanea		x						CE
Cave-dwelling Frog	Litoria cavernicola						x		DD
Red-eyed Tree Frog	Litoria chloris	x	x						LC
Blue Mountains Tree Frog	Litoria citropa		x	x					LC
Cooloola Sedge Frog	Litoria cooloolensis	x							EN

Common English Name	Scientific Name	Qld	NSW	Vic	Tas	SA	WA	NT	IUCN
Copland's Rock Frog	Litoria coplandi	x					x	x	LC
Spotted-thighed Frog	Litoria cyclorhyncha						x		LC
Dahl's Aquatic Frog	Litoria dahlii	x					x	x	LC
Davies' Tree Frog	Litoria daviesae		x						VU
Australian Lace Lid	Litoria dayi	x							EN
Bleating Tree Frog	Litoria dentata	x	x						LC
Buzzing Tree Frog	Litoria electrica	x							LC
Growling Tree Frog	Litoria eucnemis	x							LC
Brown Tree Frog	Litoria ewingii		x	x	x	x			LC
Dwarf Tree Frog	Litoria fallax	x	x	x					LC
Wallum Rocket Frog	Litoria freycineti	x	x						VU
Centralian Tree Frog	Litoria gilleni							x	LC
Dainty Green Tree Frog	Litoria gracilenta	x	x						LC
Peter's Frog	Litoria inermis	x					x	x	LC
Giant Tree Frog	Litoria infrafrenata	x							LC
Jervis Bay Tree Frog	Litoria jervisiensis		x	x					LC
Jungguy Tree Frog	Litoria jungguy	x							LC
Kroombit Tree Frog	Litoria kroombitensis	x							CE
Broad-palmed Frog	Litoria latopalmata	x	x			x			LC
Lesueur's Frog	Litoria lesueuri		x	x					LC
Northern Heath Frog	Litoria littlejohni		x						LC
Long-snouted Frog	Litoria longirostris	x							LC
Armoured Frog	Litoria lorica	x							CE
Rockhole Frog	Litoria meiriana						x	x	LC
Javelin Frog	Litoria microbelos	x					x	x	LC
Motorbike Frog	Litoria moorei						x		LC
Kuranda Tree Frog	Litoria myola	x							CE
Torrent Tree Frog	Litoria nannotis	x							EN
Rocket Frog	Litoria nasuta	x	x				x	x	LC
Bridle Frog	Litoria nigrofrenata	x							LC
Narrow-fringed Frog	Litoria nudidigita		x	x					LC
Mountain Mist Frog	Litoria nyakalensis	x							CE
Wallum Sedge Frog	Litoria olongburensis	x	x						VU
Pale Rocket Frog	Litoria pallida	x					x	x	LC
Victorian Frog	Litoria paraewingi		x	x					LC
Cascade Tree Frog	Litoria pearsoniana	x	x						NT

Common English Name	Scientific Name	Qld	NSW	Vic	Tas	SA	WA	NT	IUCN
Peron's Tree Frog	Litoria peronii	x	x	x		x			LC
Masked Frog	Litoria personata							x	LC
Leaf-green Tree Frog	Litoria phyllochroa		x						LC
Peppered Frog	Litoria piperata		x						CE
Screaming Tree Frog	Litoria quiritatus		x						LC
Riverina Bell Frog	Litoria raniformis raniformis		x	x		x			EN
Southern Bell Frog	Litoria raniformis major		x	x	x	x			EN
Whirring Tree Frog	Litoria revelata	x	x						LC
Common Mist Frog	Litoria rheocola	x							EN
Roth's Tree Frog	Litoria rothii	x					x	x	LC
Desert Tree Frog	Litoria rubella	x	x			x	x	x	LC
Green-eyed Tree Frog	Litoria serrata	x							VU
Spotted Tree Frog	Litoria spenceri		x	x					CE
Northern Creek Frog	Litoria spaldingi	x						x	NA
Magnificent Tree Frog	Litoria splendida						x	x	LC
Chattering Rock Frog	Litoria staccato						x		LC
Glandular Frog	Litoria subglandulosa	x	x						VU
Tornier's Frog	Litoria tornieri	x					x	x	LC
Tyler's Tree Frog	Litoria tyleri	x	x						LC
Alpine Tree Frog	Litoria verreauxii alpina		x	x					EN
Whistling Tree Frog	Litoria verreauxii verreauxii	x	x	x					LC
Wotjulum Frog	Litoria watjulumensis						x	x	LC
Southern Heath Frog	Litoria watsoni		x	x					EN
Wilcox's Frog	Litoria wilcoxii	x	x						LC
Orange-thighed Frog	Litoria xanthomera	x							LC
True Frogs									
Water Frog	Papurana daemeli	x						x	LC
True Toads									
South-east Asian Toad	Duttaphrynas melanostictus	x	x	x			x	x	LC
Cane Toad	Rhinella marina	x	x			x	x	x	LC
Caudates									
Smooth Newt	Lissotriton vulgaris			x					LC

Common Name	Scientific Name	Generic name applicable following Duellman et al., with correction by Dubois and Fretey	Generic name applicable following the phylogeny in Bell et al.
Striped Burrowing Frog	Cyclorana alboguttata	Ranoidea	Cyclorana
Giant Frog	Cyclorana australis	Ranoidea	Cyclorana
Short-footed Frog	Cyclorana brevipes	Ranoidea	Cyclorana
Hidden-ear Frog	Cyclorana cryptotis	Ranoidea	Cyclorana
Knife-footed Frog	Cyclorana cultripes	Ranoidea	Cyclorana
Long-footed Frog	Cyclorana longipes	Ranoidea	Cyclorana
Daly Waters Frog	Cyclorana maculosa	Ranoidea	Cyclorana
Main's Frog	Cyclorana maini	Ranoidea	Cyclorana
Small Frog	Cyclorana manya	Ranoidea	Cyclorana
New Holland Frog	Cyclorana novaehollandiae	Ranoidea	Cyclorana
Western Water-holding Frog	Cyclorana occidentalis	Ranoidea	Cyclorana
Eastern Water-holding Frog	Cyclorana platycephala	Ranoidea	Cyclorana
Wailing Frog	Cyclorana vagita	Ranoidea	Cyclorana
Rough Frog	Cyclorana verrucosa	Ranoidea	Cyclorana
Slender Tree Frog	Litoria adelaidensis	Litoria	Coggerdonia
Cape Melville Tree Frog	Litoria andiirrmalin	Ranoidea	Unnamed 1
Green and Golden Bell Frog	Litoria aurea	Ranoidea	Ranoidea
Kimberley Rockhole Frog	Litoria aurifera	Litoria	Mahonabatrachus
Kimberley Rocket Frog	Litoria axillaris	Litoria	Litoria
Slender Bleating Tree Frog	Litoria balatus	Litoria	Colleeneremia
Barrington Tops Tree Frog	Litoria barringtonensis	Ranoidea	Dryopsophus
Beautiful Tree Frog	Litoria bella	Ranoidea	Unnamed 5
Northern Dwarf Tree Frog	Litoria bicolor	Litoria	Drymomantis
Booroolong Frog	Litoria booroolongensis	Ranoidea	Euscelis
Green-thighed Frog	Litoria brevipalmatus	Nyctimistes	Unnamed 2
Tasmanian Tree Frog	Litoria burrowsae	Litoria	Saganura
Green Tree Frog	Litoria caerulea	Ranoidea	Pelodryas
Yellow-spotted Tree Frog	Litoria castanea	unassigned	Ranoidea
Cave-dwelling Frog	Litoria cavernicola	Ranoidea	Pelodryas
Red-eyed Tree Frog	Litoria chloris	Ranoidea	Unnamed 5
Blue Mountains Tree Frog	Litoria citropa	Ranoidea	Dryopsophus
Cooloola Sedge Frog	Litoria cooloolensis	Litoria	Drymomantis
Copland's Rock Frog	Litoria coplandi	Litoria	Litoria
Spotted-thighed Frog	Litoria cyclorhyncha	Ranoidea	Ranoidea
Dahl's Aquatic Frog	Litoria dahlii	Ranoidea	Unnamed 3
Davies' Tree Frog	Litoria daviesae	Ranoidea	Dryopsophus
Australian Lace Lid	Litoria dayi	Ranoidea	Moseleyia
Robust Bleating Tree Frog	Litoria dentata	Litoria	Colleeneremia
Buzzing Tree Frog	Litoria electrica	Litoria	Colleeneremia

Common Name	Scientific Name	Generic name applicable following Duellman *et al.*, with correction by Dubois and Fretey	Generic name applicable following the phylogeny in Bell *et al.*
Growling Tree Frog	*Litoria eucnemis*	*Ranoidea*	*Unnamed 4*
Brown Tree Frog	*Litoria ewingii*	*Litoria*	*Rawlinsonia*
Dwarf Tree Frog	*Litoria fallax*	*Litoria*	*Drymomantis*
Wallum Rocket Frog	*Litoria freycineti*	*Litoria*	*Litoria*
Centralian Tree Frog	*Litoria gilleni*	*Ranoidea*	*Pelodryas*
Dainty Green Tree Frog	*Litoria gracilenta*	*Ranoidea*	*Unnamed 5*
Peter's Frog	*Litoria inermis*	*Litoria*	*Litoria*
Giant Tree Frog	*Litoria infrafrenata*	*Nyctimistes*	*Sandyrana*
Jervis Bay Tree Frog	*Litoria jervisiensis*	*Litoria*	*Rawlinsonia*
Jungguy Tree Frog	*Litoria jungguy*	*Ranoidea*	*Euscelis*
Kroombit Tree Frog	*Litoria kroombitensis*	*Ranoidea*	*Dryopsophus*
Broad-palmed Frog	*Litoria latopalmata*	*Litoria*	*Litoria*
Lesueur's Frog	*Litoria lesueuri*	*Ranoidea*	*Euscelis*
Heath Frog	*Litoria littlejohni*	*Litoria*	*Rawlinsonia*
Long-snouted Frog	*Litoria longirostris*	*Litoria*	*Llewellynura*
Armoured Frog	*Litoria lorica*	*Ranoidea*	*Moseleyia*
Rockhole Frog	*Litoria meiriana*	*Litoria*	*Mahonabatrachus*
Javelin Frog	*Litoria microbelos*	*Litoria*	*Llewellynura*
Motorbike Frog	*Litoria moorei*	*Ranoidea*	*Ranoidea*
Kuranda Tree Frog	*Litoria myola*	*Ranoidea*	*Unnamed 4*
Torrent Tree Frog	*Litoria nannotis*	*Ranoidea*	*Moseleyia*
Rocket Frog	*Litoria nasuta*	*Litoria*	*Litoria*
Bridle Frog	*Litoria nigrofrenata*	*Litoria*	*Litoria*
Narrow-fringed Frog	*Litoria nudidigata*	*Ranoidea*	*Dryopsophus*
Mountain Mist Frog	*Litoria nyakalensis*	*Ranoidea*	*Moseleyia*
Wallum Sedge Frog	*Litoria olongburensis*	*Litoria*	*Drymomantis*
Pale Rocket Frog	*Litoria pallida*	*Litoria*	*Litoria*
Victorian Frog	*Litoria paraewingi*	*Litoria*	*Rawlinsonia*
Cascade Tree Frog	*Litoria pearsoniana*	*Ranoidea*	*Dryopsophus*
Peron's Tree Frog	*Litoria peronii*	*Litoria*	*Pengilleya*
Masked Frog	*Litoria personata*	*Litoria*	*Litoria*
Leaf-green Tree Frog	*Litoria phyllochroa*	*Ranoidea*	*Dryopsophus*
Peppered Frog	*Litoria piperata*	*Ranoidea*	*Dryopsophus*
Screaming Tree Frog	*Litoria quiritatus*	*Litoria*	*Colleeneremia*
Southern Bell Frog	*Litoria raniformis*	*Ranoidea*	*Ranoidea*
Whirring Tree Frog	*Litoria revelata*	*Litoria*	*Rawlinsonia*
Common Mist Frog	*Litoria rheocola*	*Ranoidea*	*Moseleyia*
Roth's Tree Frog	*Litoria rothii*	*Litoria*	*Pengilleya*
Desert Tree Frog	*Litoria rubella*	*Litoria*	*Colleeneremia*

Common Name	Scientific Name	Generic name applicable following Duellman et al., with correction by Dubois and Fretey	Generic name applicable following the phylogeny in Bell et al.
Green-eyed Tree Frog	*Litoria serrata*	Ranoidea	Unnamed 4
Spotted Tree Frog	*Litoria spenceri*	Ranoidea	Dryopsophus
Northern Creek Frog	*Litoria spaldingi*	Litoria	Litoria
Magnificent Tree Frog	*Litoria splendida*	Ranoidea	Pelodryas
Chattering Rock Frog	*Litoria staccato*	Litoria	Litoria
Glandular Frog	*Litoria subglandulosa*	Ranoidea	Dryopsophus
Tornier's Frog	*Litoria tornieri*	Litoria	Litoria
Tyler's Tree Frog	*Litoria tyleri*	Litoria	Pengilleyia
Alpine Tree Frog	*Litoria verreauxii alpina*	Litoria	Rawlinsonia
Whistling Tree Frog	*Litoria verreauxii verreauxii*	Litoria	Rawlinsonia
Wotjulum Frog	*Litoria watjulumensis*	Litoria	Litoria
Southern Heath Frog	*Litoria watsoni*	Litoria	Rawlinsonia
Wilcox's Frog	*Litoria wilcoxii*	Ranoidea	Euscelis
Orange-thighed Frog	*Litoria xanthomera*	Ranoidea	Unnamed 5

Further Information

WEBSITES

Amphibian Research Centre www.frogs.org.au
AmhibiaWeb www.amphibiaweb.org
Australian Faunal Directory www.biodiversity.org.au
Australian Museum www.australianmuseum.net.au
Canberra Nature Map www.canberra.naturemapr.org/Community/CategoryGuide/12
Frog and Tadpole Study Group of NSW www.fats.org.au
Frog ID www.frogid.net.au
FrogWatch (NT) www.frogwatch.org.au
Frogwatch SA www.frogwatchsa.com.au
Ginninderra Catchment Group www.ginninderralandcare.org.au/frogwatch
Museum and Art Gallery of the Northern Territory www.magnt.net.au
Museums Victoria www.museumsvictoria.com.au
Nature 4 You www.wildlifedemonstrations.com
Parks & Wildlife Service Tasmania www.parks.tas.gov.au/index.aspx?base=3060
Peter Rowland Photographer & Writer www.prpw.com.au
Queensland Museum (Frogs)
www.qm.qld.gov.au/Find+out+about/Animals+of+Queensland/Frogs
South Australian Museum www.samuseum.sa.gov.au
Tasmanian Museum and Art Gallery www.tmag.tas.gov.au
Western Australian Museum (Alcoa Frog Watch) www.museum.wa.gov.au/explore/frogwatch

REFERENCES

Anstis, M. (2002) *Tadpoles of South-eastern Australia: A Guide with Keys.* New Holland, Sydney.

Anstis, M. (2017) *Tadpoles and Frogs of Australia* (2nd edn). New Holland Publishing, Sydney.

Barker, J., Grigg, G. & Tyler, M. (1995) *A Field Guide to Australian Frogs.* Surrey Beatty, Chipping Norton, NSW.

Bell, R. C., Webster, G. N. & Whiting, M. J. (2017) Breeding biology and the evolution of dynamic sexual dichromatism in frogs. *Journal of Evolutionary Biology*, 30 (12): 2104–2115.

Berger, L., Speare, R. & Hyaty, L. (1999) Chytrid fungi and amphibian declines: overview, implications and future directions, in *Declines and Disappearances of Australian Frogs* (Alastair Campbell, ed.). Biodiversity Group Environment Australia, Canberra.

Catullo, R. A. & Keogh, J. S. (2021) Seen only once: an evolutionarily distinct species of Toadlet (*Uperoleia*: Myobatrachidae) from the Wessel Islands of northern Australia. *Zootaxa*, 5057 (1): 052–068.

Cogger, H. (2018) *Reptiles and Amphibians of Australia* (7th edn). CSIRO Publishing, Collingwood.

Donnellan, S. C., Catalano, S. R., Pederson, S., Mitchell, K. J., Suhendran, A., Price, L. C., Doughty, P. & Richards, S. J. (2021) A revision of the *Litoria watjulmensis* (Anura: Pelodryadidae) group from the Australian monsoonal tropics including the resurrection of *L. spaldingi*. *Zootaxa*, 4933(2): 211-240.

Dubois, A. & Frétey, T. (2016) A new nomen for a subfamily of frogs (Amphibia, Anura). *Dumerilia*, 6: 17–23.

Duellman, W. E., Marion, A. B. & Hedges, S. B. (2016) Phylogenetics, classification, and biogeography of the treefrogs (Amphibia: Anura: Arboranae). *Zootaxa*, 4104 (1): 1–109.

Eipper, S. C. (2012) *A Guide to Australian Frogs in Captivity*, Reptile Publications, Burleigh.

Hoskin, C. J., Grigg, G. C., Stewart. D. A. & Macdonald, S. L. (2015) Frogs of Australia (1.1(4614) (Mobile application software) from www. ugmedia.com.au.

Hoskin, C. & Hero, M. J. (2008) *Rainforest Frogs of the Wet Tropics, North-east Australia,* Griffith University, Gold Coast.

Littlejohn, M. J. (2003) *Frogs of Tasmania.* Fauna of Tasmania handbook No. 6, 2nd edn. University of Tasmania, Hobart.

Mahony, M. J., Hines, H. B., Bertozzi, T., Bradford, T. M., Mahony, S. V., Newell, D. A., Clarke, J. M. & Donnellan, S. C. (2022) A new species of *Philoria* (Anura: Limnodynastidae) from the uplands of the Gondwana Rainforests World Heritage Area of eastern Australia. *Zootaxa*, 5104 (2): 209–241.

Mahony, M. J., Hines, H. B., Mahony, S. V., Moses, B., Catalano, S. R., Myers. S. & Donnellan, S. C. (2021) A new hip-pocket frog from mid-eastern Australia. *Zootaxa*, 5057 (4): 451–486.

Mahony, M. J., Moses, B., Mahony, S. V., Lemckert, F. & Donnellan, S. C. (2020) A new species of frog in the *Litoria ewingii* species group (Anura: Pelodryadidae) from south-eastern Australia. *Zootaxa*, 4858 (2): 201–230.

Mahony, M. J., Penman, T., Bertozzi, Lemckert, F., Bilney, R. & Donnellan, S. C. (2021) A taxonomic revision of south-eastern Australian giant burrowing frogs (Anura: Limnodynastidae: *Heleioporus* Gray) *Zootaxa*, 5016 (4): 451–489.

Menzies, J. (2006) *The Frogs of New Guinea and Solomon Islands.* Pensoft, Moscow.

Rowland, P. & Eipper, S. C. (2018) *A Naturalist's Guide to the Dangerous Creatures of Australia.* John Beaufoy Publishing, Oxford.

Rowley, J. J. L, Mahony, M. J., Hines, H. B., Myers. S., Price, L. C., Shea, G. M. & Donnellan, S. C. (2021) Two new frog species from the *Litoria rubella* species group from eastern Australia. *Zootaxa*, 5071 (1): 001–041.

Sanders MG (2021) *Photographic Field Guide to Australian Frogs* CSIRO Publishing, Collingwood

Simon, E., Puky, M., Braun, M. & Tóthmérész, B. (2011) Frogs and toads as biological indicators in environmental assessment, in *Frogs: Biology, Ecology and Uses* (James L. Murray, ed.). Nova Science Publishers, Inc.

Somaweera, R. (2017) *A Naturalist's Guide to the Reptiles and Amphibians of Bali*. John Beaufoy Publishing, Oxford.

The IUCN Red List of Threatened Species. www.iucnredlist.org. Downloaded on 25 June 2022, 2018.

Tyler, M. J. (1999) *Australian Frogs, a Natural History*. Reed New Holland, Sydney.

Tyler, M. J. & Knight, F. (2009) *Field Guide to the Frogs of Australia*. CSIRO Publishing, Melbourne.

Tyler, M. J. & Doughty, P. (2009) *Field Guide to Frogs of Western Australia* (4th edn). Western Australian Museum, Perth.

Vanderduys, E. (2012) *Field Guide to the Frogs of Queensland*. CSIRO Publishing, Melbourne.

Webster, G. N. & Bool, I. (2022) A new genus for four myobatrachid frogs from the south Western Australian ecoregion. *Zootaxa*, 5154 (2): 127–151.

Acknowledgements

We, the authors, firstly thank our wives, Tie Eipper and Kate Rowland, for their wonderful and unconditional support and assistance during the writing of this book and our extended periods of research, both at home and in the field. We love you dearly.

Special thanks to Marion Anstis, Julian Bentley, Shane Black, Ian Bool, Brian Bush, Jessie Campbell, Patrick Campbell, Renee Catullo, Nick Clemann, Nathan Clout, Hal and Heather Cogger, Martin Cohen, Steve Donellan, Raelene Donelly, Paul Doughty, Cody Eipper, Adam Elliott, Ryan Francis, Harry Hines, Jono Hooper, Ash Horn, Conrad Hoskin, Ivy and Cory Kerewaro, Scott Keogh, Simon Maddock, Brad Maryan, Jake Meney, Michael McFadden, Angus McNab, Mark O'Shea, Adam Parsons, the late Aaron Payne, Peter Robertson, Thomas Rowland, Jodi Rowley, Mark Sanders, Glenn Shea, Jeff Streicher, Jason Sulda, Matt Summerville, Michael Swan, Riona Tindal, Janne Torkolla, Eric Vanderduys, Grant Webster, Michael Whitehead, Steve Wilson and Anders Zimny. Each of you were invaluable in many ways in the production of this book, either generously reviewing and commenting on sections of the text, giving assistance in the field, supplying the scientific resources we required for accuracy and currency of text content, providing contact details for associate researchers, or suggesting potential sources for many of the images we found difficult to obtain elsewhere.

Much of the reproductive biology information has come from the work of Marion Anstis, her work is an inspiration and provides frog workers across Australia with a wonderful resource.

We would like to specifically thank Renee Catullo for her insightful comments regarding the genus *Uperoleia* and her assistance with this challenging genus.

Lastly, we thank the publishers, including John Beaufoy and Rosemary Wilkinson, for the opportunity to write this book, series editor Krystyna Mayer for ensuring readability and consistency in text, and Sally Bird of Calidris Literary Agency for the establishment and management of these valuable relationships.